I0817597

Praise for *The Way of Excellence*

"An absolutely beautiful book that captures a lot of what I believe as a coach. There is power in committing to a process mindset, both for an individual and for a team. The gratification that comes from truly losing oneself in that process is an amazing gift. It brings joy, peace of mind, and a sense of pride. These are things we frequently talk about with the Warriors, and I believe are part of our success."

—**Steve Kerr**, nine-time NBA champion, coach of the Golden State Warriors and Team USA

"In *The Way of Excellence,* readers are invited to rethink what it truly means to be at their best—not through perfectionism or shallow success, but through purposeful growth and meaningful engagement. Drawing on science, philosophy, and stories from world-class performers, the book reveals how excellence is a deeply human process. Stulberg offers a powerful alternative to burnout culture and digital distraction, showing how anyone can cultivate a life of focus, vitality, and authentic achievement."

—**Charles Duhigg**, author of bestsellers *The Power of Habit* and *Supercommunicators*

"Brad Stulberg's examination of excellence as both a concept and a practice is food for thought for every profession and discipline—and for life itself. Weaving philosophy, science, and lived experiences into a riveting narrative, he shows us how everyone is capable of growth. *The Way of Excellence* is full of tools, but it's about more than achieving and succeeding. I loved reading it. Stulberg describes many layers of brilliance, drawing attention to the complex landscape of beautiful human contributions around us."

—**Hilary Hahn**, international violin soloist and three-time Grammy Award winner

"In a chaotic world filled with shallow distraction and performative busyness, *The Way of Excellence* is a must-read book that offers a path toward something deeper: the disciplined pursuit of mastery, competence, and mattering. Whether you're a knowledge worker, creative, or student, I highly recommend it!"

—**Cal Newport**, *New York Times* bestselling author of *Slow Productivity* and *Deep Work*

"A transformative read for aspiring high performers of all disciplines and all levels. Brad's words help me to hone my own process and ensure I'm keeping the most important things the most important things. This book is the *best* reminder that the way we do what we do matters."

—**Chelsea Sodaro,** Ironman World Champion

"*The Way of Excellence* is Brad Stulberg at his very best: relentlessly pragmatic, truly wise, and genuinely helpful. This book teaches us that in an era of shallow hacks and quick fixes, mastery requires depth and patience. Excellence is: consistency over intensity, fundamentals over fads, and progress over perfection. It is hard for me to think of anyone in my life to whom I *wouldn't* recommend this book."

—**Derek Thompson**, author of *Abundance* and *Hit Makers*

The Way of Excellence

A Guide to True Greatness and Deep Satisfaction in a Chaotic World

Brad Stulberg

HarperOne
An Imprint of HarperCollinsPublishers

Song lyrics on p. 54 by John Moreland, from "The Future Is Coming Fast," on the album *Visitor,* ℗ 2024, Old Omens/Thirty Tigers, used by permission.

HarperCollins books may be purchased for educational, business, or sales promotional use. For information, please email the Special Markets Department at SPsales@harpercollins.com.

harpercollins.com

FIRST EDITION

Designed by Jason Kayser

Library of Congress Cataloging-in-Publication Data has been applied for.

ISBN 978-0-06-338594-8

Printed in the United States of America

26 27 28 29 30 LBC 8 7 6 5 4

For Caitlin, Theo, and Lila.

I love you all.

It’s not just about the destination but the person you become along the way.

Contents

Introduction: Redefining Excellence—and Why We Need It More Than Ever

When are you at your best, and what does it feel like?

I've posed this question to hundreds of people—athletes, artists, musicians, physicians, creatives, writers, attorneys, investors, teachers, coaches, parents, and more. The answers coalesce around a few major themes:

We are at our best when we are: *Working without distraction on something that matters to us. Creating and contributing to the world. Engaging deeply with others. Sufficiently challenged. Using our unique skills. In a state of relaxed productivity. Giving something our all. In the presence of beauty.*

We feel: *Joy and inspiration. Ready and attuned. Invigorated. Electric. Nourished and at peace. Deeply satisfied. Profound awe. Fully alive. At home in the world.*

This is what excellence is all about.

Excellence is often used in reference to professional athletes, award-winning artists, and other masters of craft, so much so that people who don't fit into those categories may believe they are barred from participating in it. Other times, excellence is confused with perfectionism or winning at all costs. There's also performative excellence, which is more concerned with the appearance of being great than actual greatness. But these narratives couldn't be further from the truth.

Excellence is one of the most real things there is, and excellence is for all of us.

Throughout history, those who were constantly on the lookout for greater opportunities gained an immense survival advantage. Even if we are fortunate enough not to be worried about our next meal, that early imprinting hasn't disappeared. We have an innate drive for progress and growth. It's why we feel so alive when we channel that drive into meaningful goals—be it starting a business, writing a book, learning an instrument, studying a craft, or training for a marathon. Excellence is less a destination and more an energizing process of growth and becoming—an ongoing path that yields our best performances and, every bit as important, our best selves. We are made to move toward excellence as a tree is made to move toward the sun.

Yet when I asked the same hundreds of people how often they experience being at their best, the vast majority replied with some version of *not as often as I'd like.*

They are not alone.

The world is chaotic and full of noise. We are constantly flooded with synthetic content and experiences, much of which is dispensed not by humans but by algorithms and machines. Our attention is fractured. Our work and home lives can be stressful, often at the same time. We commonly feel distracted, overloaded, lonely, and exhausted, if not on the edge of burnout. The result is that we are pulled away from the states of mind and being that we yearn for and, eventually, from the people we aspire to become. *Alienation* describes the disconnect people experience from their own lives, and it is a defining problem of our time. When we lose the ability to pursue what matters to us with integrity, we lose a sense of who we are.

What I'm calling excellence offers an alternative.

Excellence drives purposeful attention and productive action. It supports living with intimacy and depth instead of remove and shallowness. It reconnects us with our essential humanity and fills us with purpose in an increasingly numbed-out world. Now more than ever, it is imperative that we cultivate our own versions of excellence. Our individual and collective lives depend on it. This book will show you how.

What Is Excellence?

There are times when I am deep into a book project and my reading, research, reporting, and writing—you could even say my entire being—combine to create a work that fills me with pride. Or when my athletic training gathers a momentum that nothing can stop, my body and mind syncing seamlessly to generate strength, power, and speed—and not just on any given day, but for weeks, maybe even months. During these periods, I know something good is happening. I can sense it deep inside. I feel my best, do my best, and become my best.

Perhaps you've had similar experiences.

No doubt, excellence has an ineffable quality, but it also has an empirical one, which means you can define it and map out its attainment. When you do, the result is *involved engagement in something worthwhile that supports your values and goals.*

Excellence combines mastery and mattering.

Mastery means developing skill and making progress in activities you deem worthwhile. It could result from learning a trade or advancing in a career, but it could also come from increasing the weight of your deadlift, writing poetry, building a table, making art, or learning how to play an instrument. Mattering is a sense that what you are doing has significance, that your contributions and progress are meaningful. Decades of research show mastery and mattering are key to a life well lived, or what psychologists call *life satisfaction*.

We experience the high of excellence when we are creating it, and we are drawn to it as witnesses. It's the sculpture that elicits a cascade of emotions. The motorcycle that tightly grips the road. The basketball jump shot comprised of one hundred small movements harmoniously flowing together. The craftsperson's practiced hammer strike on a nail. The rhythm of a great sentence. The leader whose calm yet forceful presence commands the attention of an entire room. The back-and-forth of an absorbing conversation. The surgeon operating at her full scope, knowing each surgical tool, incision line, and nerve ending. The poignant turn in a powerful song. Like a finely tuned instrument, excellence exudes an attractiveness and rightness that we know not only in our heads, but also in our hearts.

It's a big part of what makes life worth living.

At times you may feel at one with whatever it is you are doing, yet excellence is a longer path, a rhythm that you create and sustain. It's not just the pitcher's no-hit game, the professor's captivating lecture, or the farmer's bountiful harvest. It's also everything that went into those achievements and what is yet to come. Everyone loves to talk about the peak experiences and transcendent moments, but those moments only arise after you've been consistent with the small stuff, the day-to-day fundamentals, for some prior period of time. You've got to lay the soil and tend to it for the seeds to sprout.

Excellence

An ongoing process of growth and becoming that imbues life with meaning and vigor. It emerges from ***involved engagement in something worthwhile that supports your values and goals.***

Excellence combines ***mastery*** and ***mattering***. It is not something that is out of reach but rather it is your birthright, a core part of your nature.

Here's why this matters: When you orient around excellence, you end up getting the best out of yourself on the things that matter to you most. Nobody can buy excellence or use power, authority, or intimidation to force their way into it. Nor is excellence a disembodied, ephemeral occurrence in a virtual world. Rather, excellence is one of the most authentic and human experiences there is, and it must be earned. The payoff is a deep internal satisfaction, a genuine self-respect based on effort and competence, a sense of aliveness and resonance that you won't find anywhere else.

Excellence is also about creating a better world.

It is the generative force that leads to scientific breakthroughs, inspiring athletic feats, beautiful creations, and meaningful relationships. When you strive for excellence, you contribute to this universal dance in your own unique way.

What Excellence Is Not

There are a few common misconceptions about and impostors to the version of excellence presented in this book. Before we dive deeper into what excellence is and how to create it, it's worth being explicit about what excellence is not.

Excellence is not perfectionism.

Perfectionism is trying to do everything flawlessly. It is a trap that leads to stress and burnout. Excellence means picking and choosing what matters most, giving those pursuits your all, and letting others go. Excellence also realizes that every masterpiece begins as a rough draft, that every story of enduring progress contains highs and lows.

Excellence is not obsession.

Obsession is when you cannot stop, even when you want to. It is intrusive, all-encompassing, and enmeshes your entire self-worth with what you are doing. No doubt, excellence requires hard work. It demands commitment, persistence, and dedication. But it also includes periods of rest and renewal. It acknowledges the value of stepping away, even—and perhaps especially—when doing so is hard. Obsession results in burning bright, then burning out. Excellence is about consistency over time to create extraordinary work, and an extraordinary life.

Excellence is not optimization.

Optimization lends itself to predictability and efficiency, working like a machine. Excellence requires curiosity and exploration. It is guided by an innate knowing that resides deep in your nervous system and that's been passed down over millennia. Excellence is the opposite of mechanical. It is distinctly human, alive, and full of feeling.

Excellence is not happiness.

Happiness, at least as it's popularly conceived, involves experiencing pleasure and avoiding pain. Excellence includes a range of emotions. Even when excellence reaches its zenith, the associated feelings are more like satisfaction, vitality, and love. When I asked hundreds of people how they felt when they were at their best, nobody replied, "Happy." Some said happiness came later on, but it was never the main thing. It always emerged as a byproduct of pursuing excellence.

Excellence is not flow.

Flow describes acute moments of effortless absorption in an activity. It is values neutral, which means you can experience flow creating a masterpiece but also mindlessly scrolling social media. (The latter has been coined *shitty flow*, an idea you'll read about more in chapter 2.) Excellence, on the other hand, is values laden. It is an elongated path that requires sustained and meaningful effort. Excellence may include moments of flow, but it encompasses so much more—attributes such as caring, discipline, patience, prioritization, ritual, routine, rest, renewal, resilience, and gumption.

As you'll see in the coming pages, perfectionism, obsession, optimization, over-indexing on happiness, and becoming addicted to flow, especially the shitty variety, are some of the biggest *barriers* to excellence. It's a shame when people mistake these other qualities for excellence, thinking they are the keys to a good life, only to come up empty-handed and longing for more.

The Stakes Are High

We live in an era rife with shallow "hacks," commoditized influencers, and quick fixes. Turn on the television or log on to the internet and it won't be long before you are served up an array of false promises and performative displays of hustle-culture greatness: Do this then do that, punish yourself for small missteps, simply imagine your future and it will become reality. This stuff is pseudo-excellence: elaborate Kabuki masquerading as the real thing. Pseudo-excellence is loud, arrogant, and in your face. It cycles from fad to fad and chases attention. Real excellence is quiet, confident, consistent, respectful of the craft, and interested in the process of self-discovery.

When we fall for pseudo-excellence, or give up on excellence altogether, the costs are substantial. Data from the Gallup Polling Group show that over 75 percent of people feel burnout at some point every year, and only 32 percent feel like they are fully engaged in their work. Far too many of us report being isolated, disconnected, and depleted outside of the workplace, too. Survey research from Statista shows that 68 percent of us are "just getting by, languishing, or struggling." In all of these instances, a core problem is the diminishment of genuine excellence in our lives.

Becoming Excellent

This book aims to do two things: The first part will offer a new and comprehensive theory of excellence. It will do this by integrating findings from traditionally siloed domains. You will learn about fascinating new science showing that we are hardwired for excellence—it is an innate force that underlies sustainable progress, optimal functioning, and lasting well-being in all living species, from bacteria to humans. You'll learn how excellence drives growth and evolution, and why it's associated with deep-seated feelings of rightness and satisfaction. You'll dive into the interplay of thinking and feeling, and examine how excellence fits into a practical philosophy of life. You'll hear stories of world-class performers from across diverse domains—from music to medicine to sports to chess—and uncover what an artist can learn from an athlete, what a scientist can learn from an artist, what a craftsperson can learn from a scientist, and what they can all learn from one another. Emerging from these commonly overlooked connections in science, theory, and practice is the truth of excellence and its central importance in our lives.

The second part of this book outlines the core factors of excellence—the mindsets, habits, and practices that underlie the cultivation of excellence in oneself and others. You'll learn about the relationship between inner and outer quality, and see how both result from living and working in alignment with your values. You'll explore a new way to think about setting goals and the relationship between consistency, discipline, and intensity. You'll discover the power of gumption and the importance of staying patient and problem-solving through peaks, valleys, and plateaus. You'll see why caring deeply makes you vulnerable and why you should do it anyway. You'll learn about curiosity and confidence, rest and renewal, routine and ritual, resilience, how to design environments that support excellence in your life, and the role of milestones on what is ultimately a never-ending path. You'll also discover how to overcome common traps, pitfalls, and barriers to excellence, including alienation, distraction, chasing a never-ending supply of cheap thrills, and addiction to external validation.

If you want to jump right into the practical advice, you can start with part 2, but I recommend reading part 1 first. If I've done my job well, by the time you are finished, you will thoroughly understand what excellence is, why it's so vital, and how to build a life around it.

Part 1

FOUNDATIONS

1

The Biology of Excellence

In the summer of 2020, my family moved from Oakland, California, to Asheville, North Carolina. We rented a house on the east side of town. I wanted to keep up with my physical practice while we settled in, so I built a simple garage gym with a barbell and some weights.

I enjoyed training early in the day; it allowed me to open the door and catch a breeze before the afternoon heat rolled in. During one workout, I was completing a bent-over row, an exercise that involves hinging at the hips so that your chest is parallel to the ground, then pulling a weighted barbell to your chest. You may be thinking I am about to share a moment of excellence that gave rise to a personal best lift, but that's not at all what happened.

A fair amount of fatigue had set in by my last set. I struggled my way through the final repetitions, dropped the bar, began to stand up, lifted my gaze, and there it was: six hundred pounds of flesh, a mouthful of considerable teeth, and a tongue the size of my head, so close I could feel the warmth of its breath.

I do not have words for what came next. All I can tell you is that somehow I ended up on the stairway to the back door of the house, gazing at an enormous black bear that, for her part, was completely nonplussed. I opened the door and, with one foot inside the house and now ten feet between us, I paused and finally took stock of what happened, the bear lumbering down the driveway en route to her next adventure.

Between first seeing the bear and finding myself in the doorway, I was on autopilot. There was no thinking involved, at least not that I was aware of. I didn't sit there and contemplate, *Oh my, this situation is probably not so good for me. I had better move away!* Moments later, my cognitive mind registered what had occurred—*Holy crap! That was a bear! I could have reached out and touched it. I cannot believe that just happened.* But when I first encountered the bear, I simply reacted. It was a pre-intellectual experience guided by an inner knowing I felt in every corner of my being.

Had I experienced excellence?

In many ways, the answer is a resounding yes. It was involved engagement in something worthwhile that supported my values and goals—in this case, staying alive with my limbs intact.

Though most of us don't associate excellence with dodging bears, as you'll see in the coming pages, it is this same inner knowing, this same deeply felt sense of what is good and right, passed down over millennia of evolution, that pulls us toward greatness, even when our lives aren't on the line. It explains the attunement we feel when we enter the zone—whether it's creating music, writing poetry or prose, performing at our best during an athletic competition, drawing a picture, holding stage during a pivotal speech, or settling into a groove on a big project. It also explains the intuitive appreciation, the feeling of rightness, that accompanies bearing witness to a beautiful piece of art, a comedian about to nail his punchline, or a master chef's finest creation.

When people are asked to describe what draws them to these objects and acts, what stops them in their tracks, rarely, if ever, can they. Words do not suffice. Instead, they almost always report a visceral *feeling* of being moved. "The people who weep before my pictures are having the same religious experience I had when I painted them," remarked the abstract artist Mark Rothko. The comedian Jerry Seinfeld says that when a joke lands, "Bang. You just *feel* it. You feel it like hitting a baseball on the button."

But from where do these feelings come? And why do we have them in the first place?

In this chapter, we'll discover that our attraction to excellence and the satisfaction we gain from pursuing it traces itself to our earliest ancestors. We'll see how the powerful feelings underlying excellence are central to our biology, encoded in our DNA—and how we can reclaim and harness them for our own good. It all starts with a fascinating story, one that takes us back to the beginning of time. It demonstrates that excellence is a life force, and not just in the metaphorical sense.

Excellence as a Force of Nature

Nearly four billion years ago, out of a primordial stew of energy and molecules emerged the first life on Earth. Bacteria are composed of a single cell. They have neither body nor brain. And yet, like all other living organisms that followed, they possess a deep desire to survive and proliferate—their equivalent of flourishing. Bacteria achieve their goal by relying on what biologists call "sensing and responding," which is exactly what it sounds like: Bacteria sense whether an environment is conducive to their survival, and then respond by moving toward or away from it.

"In their un-minded existence, it turns out [bacteria] assume what can only be called a sort of *moral attitude,*" writes Antonio Damasio, a professor of neuroscience, psychology, and philosophy at the University of Southern California. They are attracted to what is "right" for their survival and turned off by what is "wrong."

From bacteria evolved the first multicellular organisms. Though their needs were more intricate, they still lacked consciousness. Nevertheless, with gradually increasing accuracy, these early organisms moved toward conditions that were advantageous to their survival (e.g., food) and away from those that were not (e.g., extreme climates).

Time passed, and life inched toward greater complexity. Around five hundred million years ago, species developed a connection between the brain and body that gave rise to subjective experience. Nowadays we call this a nervous system. At first, the nervous system allowed organisms to feel only internal processes, such as digestion and respiration. However, as evolution charged ahead, organisms began to feel features of their external environments as well, such as temperature, moisture, and airflow. Many scientists consider these feelings to be the origins of a more robust and advanced consciousness.

As living creatures became more sophisticated, they developed a feature biologists call *homeostatic upregulation*: an innate drive toward flourishing, both now and in the future, with feelings serving as the guide. A negative feeling tells an organism to back off, to make a change because something is wrong. A neutral feeling represents stability. A positive feeling is a signal to keep going. When an organism is moving toward increasing levels of homeostatic upregulation, it is, by definition, demonstrating involved engagement in something worthwhile that supports its values and goals.

What's fascinating is that much like the intrinsic sense that leads us to jolt away from bears, all of this unfolded before any of these species could generate abstract thoughts. For millions of years, the path to excellence was entirely pre-intellectual. It's the hardwiring all of us have inherited, and it remains core to who we are today.

Most of us have values and goals that extend beyond mere survival: producing creative work, forging meaningful relationships, improving as an athlete or musician, becoming a better leader, making art, or mastering a craft. But our feelings continue to play a central role in steering us toward homeostatic upregulation, toward excellence.

Anyone who has performed at their best knows firsthand what an awesome experience it is, whether it's playing a sport such as tennis, becoming absorbed in prayer or meditation, riding a wave of musical harmony, steering a great conversation, or following a hunch to a novel insight. It's true even for what we consider the most intellectual endeavors. John von Neumann, the mathematician who birthed the ideas behind the modern computer, quantum mechanics, game theory, and artificial intelligence, was known to say that he could *feel* a breakthrough idea taking shape in the back of his mind. When this feeling occurred, every cell in his body compelled him to pursue it. We cannot think our way to these generative states, and it is not our thoughts that tell us when we've arrived. The peak of human experience is a feeling.

We get out of our head and into our being. Our small, separate, and protective ego that is normally worried about failure takes a backseat. We feel at one with whatever it is we are doing because, in many ways, we are. The difference between subject and object, between us and our activity, melts away.

It's what Rothko meant when he described painting as a religious experience. It's why so many great scientists are also mystics. And it's why the most compelling truths are the ones we know not just intellectually but also in our bones.

Developing Competence

An unfortunate by-product of living in a chaotic world is that many people find themselves disconnected from their instincts and what will truly fulfill them. Whether or not we're aware of it, we all long for the pinnacle of excellence, when we are fully expressing ourselves and guided by our feelings to the next right action. But nobody attains it overnight.

Every world-class physician once struggled as a resident. Every award-winning writer once stumbled their way through sentences—and at times, likely still does. (I know I certainly do.) Even something as implicit as riding a bike requires a learning period that includes at least some degree of effortful thinking.

In most activities, you've got to *make* things happen before you can *let* them happen. And even then, nobody holds on to peak experiences forever. You cycle in and out of them.

A model of human development called *the four phases of competence* does a nice job outlining how progress unfolds in nearly any endeavor. Developed in the 1970s by an employee of the Gordon Training Institute, it shows how experts have been wrestling with the role of feelings in excellence for quite some time.

The first phase is *unconscious incompetence,* which is what you experience when you are new to something. You don't know what you're doing, and you don't know that you don't know what you're doing. At this stage, the focus ought to be on humility and learning the basics. You have to get your bearings before you can make real progress.

The second phase is *conscious incompetence,* when you are making mistakes but you are aware that you are making mistakes. During this phase, deliberate thinking and effortful trying help you to move forward.

The third phase is *conscious competence,* when you are doing your activity correctly but not without labor. You rely upon effortful thinking to process moment-by-moment feedback. Gadgets, wearable devices, and data tracking may prove useful, confirming that you are on the right track and alerting you when you are not. You are trying hard and finally getting it right, which is a wonderful feeling. Yet it is also at this point that many people get stuck. That's because in order to progress to the final phase of development, you must release from thinking and trying altogether.

The fourth and final phase is *unconscious competence,* which requires shedding the effortful trying that got you this far. Here is where you ditch the GPS watch, the heart rate monitor, the instruction manual, and the pressure you put on yourself to succeed. The various tools that helped you along the way now become barriers between yourself and the direct experience of whatever it is you are doing.

A profound example of unconscious competence comes from the three-time Grammy winner and international superstar violinist Hilary Hahn. When she described her experience onstage as a soloist, often playing in front of thousands of people, she told me she feels her way forward. "It's not really a thing I'm thinking about. I'm not thinking, *Oh, I should think about the accuracy of this particular passage,* or *Oh, how am I going to vibrate this note to make this effect?* If I start thinking about that onstage using words instead of feelings, I'm already way behind in the music. It really is that you have to be completely in the note and not behind yourself or ahead of yourself because you'll miss the moment, you'll overlook all these things that are really interesting that are happening too fast for thinking."

It's not just Hahn. As you'll see in the coming pages, all of us can benefit from getting in touch with the feeling of excellence.

Feeling Your Way to Excellence

In 2011, in the *Journal of Consciousness Studies,* the scientists Duarte Araújo and Keith Davids published a groundbreaking paper, "What Exactly Is Acquired During Skill Acquisition?" They argued that unlike in old models, where skill is described as a primarily cognitive event—something you learn intellectually and develop a mental model for—in actuality, skill is "an adaptive, functional relationship between an organism and its environment." Put differently, skill isn't just something you acquire; rather, it's the ability to sense and respond, to feel your way through a complex environment. In the words of Stuart McMillan, a coach to more than thirty-five Olympic medalists in sprinting and other power sports: "Skill is not an act or action, it is an interaction—between a person, their activity, and their environment."

Research shows that the best way to learn something is to feel what it's like to do it incorrectly and correctly. You *feel* the perfect golf swing, swim stroke, running stride, or tennis serve. You *feel* the keys on the piano or the strings on the guitar. You *feel* the crowd's response to a speech or presentation. You *feel* when the painting or song is just right.

"When I'm working on something, obviously it's not going to feel exactly what it looks like," explains Tiger Woods, arguably the greatest golfer of all time. "But it gets to that point when feel and real start intersecting, when they start feeling like they're the same, that's when some pretty good magic starts to happen."

A hallmark of excellence is feeling your way into a rhythm.

Sometimes that rhythm is acute, as in the case of a golf swing or musical note. But it can also be longer lasting, such as when you gain a feel for a larger project, a sport, a leadership role, a relationship, a line of academic inquiry, or even an entire phase of life. In each of these instances, trying to think, or worse yet, force, your way into the right actions almost always backfires. You get stiff, rigid, and underperform—or even choke.

Alex Honnold is someone who cannot afford to choke.

Known for his starring role in the Academy Award–winning documentary *Free Solo,* Honnold is considered one of the greatest rock climbers in history, scaling nature's most challenging walls without any support—literally without a rope. If he falls, he almost certainly dies. There's not much else that requires that level of sustained concentration, where the downside of even the smallest misstep, error, or lapse in focus is catastrophic.

When I spoke with Honnold, he told me that when he's preparing for a challenging ascent, he spends time "visualizing what the entire experience will feel like, what specific sections will feel like, and even what individual moves will feel like."

Honnold knows that when he's on the rock face of a towering mountain, thinking too much will get him into trouble, and with perilous consequences. So he practices by acquainting himself with how the climb ought to *feel.* When he's actually out on the mountain climbing, he feels his way in and out of certain positions, continually adapting to the conditions around him.

Honnold's experience mirrors that of Tiger Woods and Hilary Hahn. But it isn't specific to sports or music. Surgeons, courtroom lawyers, traders, sculptors, coaches, and educators have all told me some version of the same thing: Thinking may be a significant part of what they do, but when they are at their best, it is their feelings that take center stage.

When I'm in a writing groove, be it on a specific day or over the course of an entire project, I don't think my way forward as much as I feel my way forward. Likewise, the award-winning author George Saunders says his edits are guided by an inner knowing: "The subconscious *feels* things, it *prefers* things; it finds *this more fun than that*... and maybe it doesn't know why. And I find it's best to say, at such moments, 'As you wish.'"

The sociologist Richard Sennett, who is widely known for his work studying craftspeople, coined the term *situated cognition* to describe how the better someone gets at an activity, the less they think with their head and the more they think (and feel) with their entire being. "When we focus on making a physical object, or playing a musical instrument, our concentration level is mainly self-directed," he explains. Generally speaking, once we reach a certain level of proficiency, if we spend too much time intellectualizing, our performance and experience deteriorates. We perform best when we feel our way forward, experiencing an intimate and involved absorption with our activity. It's an advanced version of sensing and responding, bestowed upon us from bacteria and refined over millennia of evolution. Our nervous systems evolved for the pursuit of excellence, which explains why it's such a powerful antidote to the dysregulation so many of us experience today.

When we remove distractions, focus intently, and pursue excellence, we become situated in ourselves and situated in the world.

In a fascinating series of studies, the neuroscientist Antonio Damasio and his colleagues investigated what happens when people suffer damage to their ventromedial prefrontal cortex (vmPFC), a center of feeling and emotional processing in the brain. These patients had undergone strokes or other traumatic events that caused injury to the vmPFC while sparing other neural networks. Most notably, they had no damage to regions associated with rational processing. As such, when they were given strict tests of intellect—for example, IQ tests, math problems, logic puzzles, or memory games—they had few issues. But when it came to navigating and making decisions in the real world, they were incapable of functioning at even the most basic levels. They could not operate in any environments outside of those that were fully controlled.

To explain this, Damasio and his colleagues put forth the "somatic marker" theory: without emotional assignments of value, we become incapable of making real-world decisions. When we are operating under uncertainty, when there are numerous possibilities for what action to take, it is nearly impossible to deduce whether something is "good" or "bad" without somatic markers, without the *feelings* that accompany our actions and reactions as we go about the world.

Damasio's experiments represent an extreme case, but they highlight an underlying truth about our nature. We are at our best when we get into the zone, function in lockstep with our surroundings, and feel our way toward excellence. These deeply satisfying experiences fill our lives with meaning. When we lose our ability to feel, we struggle mightily. Either way, our feelings are indispensable to our success. They guide us toward what biology calls homeostatic upregulation and what I call excellence.

Our Deepest Nature

The main reason I write and lift weights is because a technology company can't design a robot that can make me feel what I feel when I write and lift weights. Maybe the robot can write more elegantly or lift more weight. But it can't make me feel the aliveness of a great idea. It can't make me feel the rhythm of a great sentence. It can't make me feel the heavy weight starting to move. This feeling is what makes us human.

Find it in your own life. Protect it. Cherish it.

In his book *The Creative Act,* music producer Rick Rubin describes a version of excellence that he refers to as the *ecstatic*: "Following the natural feedback in the body, we move closer toward the option that hints at the ecstatic. [It is] our compass, pointing to our true north. The feeling is an affirmation that you're on the right path . . . a visceral, body-centered reaction, not a cerebral one. . . . Of all the experiences that occur during the creative process, touching the ecstatic and allowing it to guide our hand is the most profound and precious."

I hope this chapter has helped you understand the mechanisms behind Rubin's ecstatic, behind the incredible feelings that accompany the pursuit of excellence. I also hope it's helped you realize that excellence is one of the truest, most biologically attuned ways of knowing there is. Excellence is in our nature, dating back billions of years. We evolved to strive for excellence, which is why it ushers us toward our greatest creations and fills us with satisfaction.

It's also why too much automation, mechanization, distraction, or anything else that comes between us and what we care about—in essence, alienating us from our own lives—leaves us numb, empty, and burned-out.

There is, however, a trap.

It is common, insidious, and can jolt us off the path in an instant.

As indispensable as our feelings are to excellence, sometimes they are misguided. For instance, an activity or pursuit may feel wonderful in the moment but yield deleterious consequences later on. This is where our psychology and thinking minds come into play. We'll turn to both in the next chapter.

Chapter Summary

- Excellence means involved engagement in something worthwhile that supports your values and goals. It combines mastery and mattering. It is not just for athletes, artists, or other masters of craft—it is for all of us, what we are made to do.
- Scientists call our innate drive toward excellence *homeostatic upregulation*, and it traces itself back to the beginning of life. In the past it dealt mainly with survival, but now it pushes us toward flourishing.
- Progress advances through four phases of competence: unconscious incompetence, conscious incompetence, conscious competence, and unconscious competence.
- Feelings are indispensable to excellence—they play a significant role in telling us when we are on the right track: We don't think our way to excellence as much as we feel our way to excellence.

We are hardwired for excellence.

It situates us in ourselves and the world.

2

The Psychology of Excellence

Say you have a few hours on a Saturday afternoon. You can either focus on something that matters to you and that you've been wanting to do—writing a book, connecting with your partner, playing with your child, creating stained glass, or going to the gym—or you can mindlessly scroll social media, check and recheck your email, and tap away at *Candy Crush*.

In the short term, getting lost in your phone might feel good. It requires little to no effort, and it satisfies your innate drives for excitement and novelty. In the exhaustion and chaos of the world, sometimes it's just plain old comforting to turn off your brain and zone out.

In the long term, however, zoning out on your phone, especially if it becomes a repeated habit, is likely unconducive to your values and goals. At the end of the day, you probably won't feel great. You'll know that you wasted an opportunity to do something meaningful.

Many people even report a psychological hangover following these episodes. It's a classic case of acute feelings being in conflict with their chronic consequences.

A best practice is to think about what getting sucked into your phone would *feel* like after the fact. The answer, for most of us, is lousy. Getting in touch with that future feeling helps you make the right decision in the present moment.

Our ability to think, and therefore feel, our way into the future is a big part of what separates us from other species. In this chapter, we'll learn how this ability is essential to not only making it through the day but also the pursuit of excellence. In turn, we'll continue to explore how the pursuit of excellence reverses disconnection, returns us to ourselves, and makes our lives richer, more textured, and less fragmented. We'll also further examine the relationships between excellence and flow, excellence and happiness, excellence and stress, and excellence and meaning. Finally, we'll learn about a key psychological concept that helps lay the groundwork for more excellence in our lives.

Thinking and Feeling into the Future

Navigating the world based solely on direct experience is costly. For example, the first time a snake bites you, it is painful, and there is a strong negative feeling associated with the snake. If you are lucky enough to survive the bite but fail to form a conceptual idea of snakes, the next time you see one, you may touch it to see what happens. Eventually, however, your luck will run out, and you'll suffer a lethal bite. Over time, organisms that could *imagine* what a snake bite (along with all manner of other harms) might feel like gained an immense advantage over organisms that relied solely on direct experience. The more complex and accurate a species' ability to form conceptual ideas about the world, the better their chances of survival. This is a leading theory as to how thinking evolved.

"Could you imagine what would happen if each of us had to learn from experience what happens when you jump off a cliff or pick up a viper?" writes the neuroscientist Mark Solms in his book *The Hidden Spring: A Journey to the Source of Consciousness*. Thinking allows us to simulate how an experience might feel. In doing so, we can appreciate its potential benefits and harms without putting ourselves at risk.

As cognition advanced, our earliest human ancestors were able to consider not only immediate but also future consequences of their actions. This is important because it's not always adaptive to go along with our initial feelings and impulses. Millennia ago, intense fear may have manifested when coming face to face with a tiger, or intense excitement when met with the chance to reproduce and pass on our DNA. The response to those feelings would have been simple: In the first instance, get out of Dodge; in the second, procreate. Today, however, we live amid myriad paper tigers and shallow sources of excitement.

Many of us, perhaps without even fully recognizing it, are craving the real, deeply felt experience of excellence toward which our nature predisposes us. But with countless superficial substitutes on offer—not to mention ever-shifting goalposts for so-called success and happiness—we all too easily find ourselves overwhelmed and fatigued while bored, dull, and wanting for something more. Much of modern life is characterized by a restless exhaustion, and it manifests in surprising ways.

In the late two-thousands-teens, the social scientist Michael Bang Petersen and his colleagues coined a concept to describe intensifying political and social unrest, along with the proliferation of ever more wild conspiracy theories. They called it *The Need for Chaos*. It explains how we desperately want to feel alive and excited. But without more enriching options, that desperation reveals itself in people actively supporting—and in some cases, directly initiating—chaos, be it by electing polarizing political leaders, spreading ridiculous conspiracies, engaging in destructive behaviors, cheering on damage to public spaces, or supporting the destruction of civil institutions, norms, and decency. A 2024 meta-analysis in the journal *Nature Communications* found the factor showing the strongest association with conspiracy beliefs to be "social alienation." Additional research shows that boredom is also a driving force.

If people are numbed out by a lack of genuine connection, excitement, and dynamism, even disorder and mayhem start to seem enticing. They become so starved of authentic emotional resonance that they'll accept just about anything that makes them feel. It explains why even the best of us get caught up in doomscrolling, gossip, political outrage, and checking and re-checking websites and news alerts on our phones. We crowd out any stillness and space for our minds. We do anything to fill the void, even if the result is a groggy, worn-out feeling. But this is absurd, and it doesn't have to be this way. There are deeper, more fulfilling, and more productive avenues to feeling alive and reconnecting with ourselves. These deeper and more fulfilling avenues may not be the default, but that doesn't mean we can't choose them.

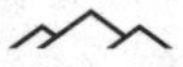

It's not just doomscrolling and chaos; there are many situations in which our short-term feelings—whether temptation, fear, or excitement—conflict with our values and long-term goals. It's why people who struggle with binge eating disorders are taught to pause and think about how they'll feel not while gorging on chocolate now but rather when they wake up the next morning. If they can learn to tap into this future feeling, it subjugates their drive to overeat. It's good counsel for any temptation, from having an extra drink to saying something to a colleague you might regret. If you can think ahead, which in turn enables you to feel ahead, you give yourself a better chance of making the right decision in the moment.

When it comes to the pursuit of excellence, we are often faced with the comfort, pleasure, or ease of short-term decisions that contradict our values and long-term goals. Anyone who has been tempted by staying in a warm and cozy bed instead of going for that brisk morning run knows this firsthand. The same is true when we are enticed by hacks, quick fixes, and shortcuts instead of committing to sustained hard work. We live in a world that encourages superficial seeking, yet it is unwavering commitment—including enduring some downright drudgery—that gives rise to genuine excellence. The key is to think beyond your immediate feelings and check them against your desires for the future.

Psychologists use the term *episodic future thinking*, or EFT, to describe the ability to imagine different futures and what they'll feel like. If we can connect these imagined futures to our values and goals, we give ourselves the best chance to stay on the path.

Selecting Worthwhile Pursuits

Decades of research on human motivation, satisfaction, and fulfillment demonstrate that we thrive over the long haul when three core needs are met: (1) *autonomy*, or the ability to have some control over how we spend our time and energy; (2) *competence*, or a path toward concrete improvement in our chosen pursuits; and (3) *belonging*, or a sense of connection to something beyond ourselves—be it another person, community, lineage, or tradition.

The more time and energy we spend on pursuits that afford us autonomy, competence, and belonging, the better.

When evaluating career opportunities, projects, and goals, we can ask ourselves, *Will it increase or decrease autonomy, competence, and belonging in my life?*

For existing projects, we could consider if there are actions to enrich our core needs. How can we protect our autonomy? If we are working with others, what conversations might help? What can we do to set worthwhile goals? How can we stay connected to something larger than ourselves? What would it look like to shape our lives for more mastery and mattering?

Excellence also asks that we spend our time and energy in alignment with our values.

Our values represent our guiding principles, the attributes and qualities that matter to us most. A few examples include *truth*, *presence*, *health*, *community*, *spirituality*, *relationships*, *intellect*, *creativity*, *quality*, *craft*, *authenticity*, and *wisdom*.

Most people have three to five core values. If you are unsure of your values and don't know where to begin, think of people you respect and then ask yourself what it is about them that you admire. You could also imagine yourself ten, twenty, or maybe even thirty years down the road, looking back on current you. What qualities would older and wiser you be proud of? Both of these strategies are inroads to uncovering your values.*

A simple yet powerful litmus test is to ask yourself, *Does this project support my values?* If you value health but your situation leaves you no time to exercise or sleep, the answer is no. If you value creativity and spend time painting or songwriting, then the answer is yes. At times, your values may come into conflict with one another. That's okay. It forces you to consciously evaluate trade-offs instead of being blindly pulled along by inertia. True success is living a life that is in alignment with your values. If you don't define your own version of success, someone else will define it for you.

Even seemingly trivial activities, such as deadlifting a certain amount of weight or running a fast mile, can be filled with meaning insofar as they align with your values. That's because when you are working on an activity, that activity is also working on you.

Reflecting on his experience as a craftsperson, Peter Korn writes, "I was engaged in furniture making as a creative process, the practice of which would help me to forge a good life. . . . For the past decade I had been imagining that my goal was to make furniture that expressed certain values. Now I saw that what I had really wanted all along was to cultivate these same qualities within myself."

Remember: Excellence is not a destination; it is a process of becoming. The real reward isn't a bigger deadlift, a faster mile, or a sturdier table; it's that you become a better version of yourself.

The world champion mountain biker Kate Courtney once told me that her effort and dedication, the hours upon hours she spends in the saddle, is as much, if not more, about inner transformation

* If you'd like guidance in selecting and defining your core values, there is an exercise in appendix 2 to help.

as it is about winning medals. "The ultimate goal is to learn about myself and use the experiences of competition to better myself not only as an athlete but as a person."

Four-time Olympic gold medalist and two-time NBA champion Kevin Durant says, "I can't walk around with my gold medals on, but the memories of all the experiences and the journey—the plane rides and bus rides and practices—I can carry that around with me every day. But medals, I'm not about to have that in my pocket or around my neck every day. But I can walk around with all of this experience and all of these stories and all of this knowledge I got from [competing at the highest level] and that can truly help somebody."

Korn, Courtney, and Durant are describing the impact excellence has on their *character*, which comes from the Greek *charassein*, meaning "to engrave or stamp upon." When we throw ourselves into worthwhile projects and pursuits, we engrave or stamp upon ourselves the type of person we are growing into.

Excellence and Flow

Focusing on worthwhile pursuits is a crucial factor that distinguishes excellence from the popular psychological construct *flow*. Flow is values neutral and agnostic to the future. You can achieve a flow state by creating a masterwork or bonding with a loved one, but you can also achieve a flow state by gambling or arguing with a stranger on the internet. Excellence, on the other hand, is values-laden—it must be pointed toward something that matters to you and supports your goals.

The psychologists David Pizarro and Paul Bloom coined the term *shitty flow* to describe the experience of being in flow but in ways that do not encourage or, even worse, work against your values and goals. Shitty flow is particularly common "when your days are filled with bullshit like emails and administrative meetings and then you're just tired, and this is the time when you should be doing something [meaningful], but all you want to do is sit and browse Reddit or maybe YouTube on your phone," explains Pizarro.

During these experiences, you may feel relief or calm and undergo some hallmarks of flow, such as losing a sense of time and perhaps even losing a sense of yourself. But when you come out of them, you are left with the sobering reality that your time could have been better spent. The result, Pizarro and Bloom observe, is often shame or disgust: "Shitty flow doesn't give you a feeling of meaning. You're not fully engaged in body and mind. It's a trick."

A particularly stark example of shitty flow comes from the anthropologist Natasha Dow Schüll, who studied people playing slot machines in Las Vegas. She found the stickiness of gambling addiction is less about the potential for winning money and more about the trancelike state gamblers achieve while using the machines. She coined the term *machine zone* to describe the experience of daily worries, social demands, and even bodily awareness slipping away. None of this happens by accident. Schüll found that slot-machine designers know precisely what they are doing—engineering addictive flow experiences.

Another crucial difference between excellence and flow is that the latter is traditionally understood as effortless, whereas the former almost always includes at least some exertion. In many cases, it's a matter of scale. Flow concerns itself with what is happening in the moment, but excellence sweeps across a broader horizon; it's about sustained effort, a groove you assemble over time. You'll almost certainly enter "the zone," but there will also be periods that you find hard and mundane. It's impossible to be in flow all the time, and chasing it will leave you frustrated and wanting; with excellence, it's the *totality* of the path, including the resolve you build in valleys and on plateaus, that leads to its enduring satisfaction.

Excellence and Happiness

Dating back to the ancient Greek empire, happiness—what Aristotle called "the highest good . . . the ultimate end"—has been a primary goal for most people. So it's worth asking: Does the way we spend our time and energy make us happy? Are there things we can do to become even happier?

The word *happiness* itself has long been a subject of debate. Modern psychology pits eudaimonic happiness, which deals with finding meaning and striving for self-realization, against hedonic happiness, or attaining pleasure and avoiding pain. Whereas hedonic happiness is often passive, eudaimonic happiness requires effort. It's the feeling you get after a long day's work on a worthwhile project. Research shows people report better in-the-moment feelings with hedonic activities, but lasting satisfaction comes from eudaimonic ones.

When you pursue something solely for pleasure in the moment, you may achieve an intense high, but you're later left with a feeling of emptiness and longing. Meaning and self-realization require more effort up front, but the positive feelings they generate are enduring.

A famous thought experiment in philosophy gives you the option to choose from two alternatives: You can enter into a tube and never leave it for the rest of your life, but while you're in the tube, you'd be injected with a steady stream of drugs that make you forget that you're in a tube and feel euphoric all the time. Or you could continue living your life, with all its hardships, hopes, and uncertainties for the future. Almost nobody chooses the tube.

Life is hard. It's not supposed to be perfect. Long-term satisfaction always includes short-term challenges. Almost everything worthwhile requires effort. Perhaps happiness, particularly in a culture saturated with the superficial variety, is the wrong goal altogether. "This pursuit of happiness seemed to have become like the pursuit of some scientifically created mechanical rabbit that moves ahead at whatever speed it is pursued," writes Robert M. Pirsig in his book *Lila: An Inquiry into Morals.* "If you ever did catch it for a few moments it had a peculiar synthetic, technological taste that made the whole pursuit seem senseless."

A better alternative to obsessing over happiness is focusing on developing skills in worthwhile pursuits while opening your emotional aperture to a range of feelings along the way. When I am deep into a book project, at week nine of a challenging twelve-week training cycle in the gym, or engrossed in coaching my son's basketball team, I wouldn't say I am *happy*. These endeavors require real work. Yet there are few things I'd rather be doing and few feelings I'd rather have. These pursuits give my life texture. What I am is fulfilled.

Excellence and Stress

In the middle of the twentieth century, the Hungarian Canadian endocrinologist Hans Selye developed what to this day remains the leading theory of stress. When you undertake demanding work or experience an acute challenge, your body produces inflammatory proteins and the hormone cortisol. These molecules serve as messengers to the rest of your system, in essence proclaiming, *This is a significant stimulus that will require a significant response.* As a result, the body marshals the appropriate resources and directs them to the area under stress. If the dose of stress isn't too great, you build back stronger and become more resilient. This represents a healthy stress response. If, however, the amount of stress is too large or lasts too long, then the body fails to adapt and the result is deterioration. Selye called this the *exhaustion stage*. Today, many refer to it as being under chronic stress, nervous system dysregulation, or burnout.

In the exhaustion phase, the body enters a catabolic process, or a state of persistent breakdown. Rather than acting as a signal for repair and then subsiding, inflammation and cortisol remain elevated. The adrenal system, constantly on edge, becomes fatigued. It's why chronic stress contributes to numerous health problems—a living system can only withstand so much tension before it breaks. "If we are doing too much," explains Selye in his 1956 book *The Stress of Life*, "the great remedy here is to learn to relax as quickly and completely as possible."

In my prior work, I boiled down Selye's model to a simple equation: *Stress plus rest equals growth*. Too much stress, not enough rest and the result is injury, illness, fatigue, and burnout. Too much rest, not enough stress and the result is complacency or stagnation. Progress and growth in nearly every domain relies upon finding the right amount of stress and following it with the right amount of rest.

However, in Selye's later and lesser-known research, he found there is an additional layer to stress, and a crucial one at that. As he aged, he increasingly incorporated psychological insights into his studies. He recognized that our response to stress is contingent not just on the dose but also on what, if any, purpose the stress serves.

Effort pointed toward a goal we find meaningful will elicit an entirely different stress response than the same effort pointed toward a goal we find meaningless. We can burn ourselves out by doing empty work and getting caught up in the rat race, or we can find fulfillment by exerting ourselves in worthwhile activities, all while working the same number of hours. It's the difference between a soul-sucking job you do only for the money and a vocation that aligns with your values and goals.

In other words, stress is mediated by meaning.

"Man must have some goal, some purpose in life that he can respect and be proud to work for," Selye writes in his 1974 book, *Stress without Distress*.

Modern scientists have built upon Selye's hypothesis and overwhelmingly demonstrated that stress is, in fact, contingent on its purpose. Studies show that our response to challenge—all the way down to the hormones and inflammatory proteins we secrete—is directly related to whether or not we view a challenge as worthwhile.

In 2022, a team of neuroscientists at McGill University in Montreal showed that when we apply effort on challenging tasks, it boosts activity in regions of the brain that respond to rewards. Essentially, they looked under the hood and found the neural networks underlying the contentment and satisfaction we feel after a hard day's work on something we care about. But here's the catch: The neural activity associated with reward, along with the accompanying real-life feelings of accomplishment, was significantly greater in participants who viewed their effort as worthwhile. In layperson's terms, working hard on something you find meaningful, even if it is stressful at times, rewires your brain in ways that lead to satisfaction and fulfillment. The more you believe your hard work is rewarding and satisfying, the more rewarding and satisfying you'll find it.

Long before we had technology to monitor brain activity, Selye observed the same connection between satisfaction and worthwhile exertion in not only his research participants but also himself. He was a devoted scientist, working for decades upon decades, navigating highs, lows, frustrations, breakthroughs, and everything in between. Yet he did not burn out, languish, or succumb to exhaustion. He attributes his longevity to finding "the capacity to contemplate the harmonious elegance of nature, at least with some degree of understanding, one of the most satisfactory experiences of which man is capable." The happiness and satisfaction he experienced as a by-product, what he described as "an equanimity and peace of mind which can be achieved only through contact with the sublime," was anything but shallow or synthetic. It was deep and real.

It's time for a new concept: *zombie burnout*.

It represents the half-dead, half-alive shuffling through your day that leaves you both restless and exhausted at the same time. Yes, you can burn out from doing too much. But you can also burn out from not doing *enough* of what lights you up.

Still, so many of us end up chasing goals we don't really care about. We follow the crowd, mimic metrics, and say yes to things that look good on paper. But imitation without meaning breeds fatigue. Zombie burnout doesn't crash your system, it dulls it. The way to treat and prevent zombie burnout is to identify what numbs and distracts you, and then do your best to trade it in for *meaningful* activities and challenges.

You don't need to be a world-class athlete, award-winning

musician, or artist featured at the Louvre. (Though you don't know if you can be until you try.) Rather, all of us can reap the benefits of excellence and contribute to the world by reconnecting with our unique gifts, talents, and values, and channeling them as best we can. If this feels overwhelming right now, part 2 of this book will show you how.

Our theory on the foundations of excellence is coming together. We are built for involved engagement in worthwhile activities that support our values and goals. Unlike happiness or flow, excellence is a longer and more textured path. The most common pitfall on that path, the one that knocks nearly everyone off, is when we spend too much time and energy on superficial and synthetic pleasures at the expense of activities that yield deeper and more lasting satisfaction. Fortunately we are equipped with episodic future thinking (EFT) systems that allow us to simulate diverse futures, helping us to stay pointed toward our true north.

Another way to think of it is that excellence relies on an ongoing back-and-forth between our visceral, innate knowing (biology) and our cerebral, explicit thinking (psychology). The former guides the process. The latter is a necessary corrective.

Overcoming Dysevolution

We did not evolve to live amid potato chips, French fries, TikTok, Pornhub, "X" (formerly known as Twitter), and online gambling. These technologies are designed to prey upon our primal instincts for food, sex, status, and connection without ever truly satisfying us. The entire business model is predicated on making us feel great for a short period of time and then leaving us wanting—so that we quickly come back for more. It's the same reason people become addicted to drugs: A powerful high combined with a short half-life creates an overwhelming hook.

For over 99 percent of our species' history, we lived in scarcity. Abundance is a recent phenomenon. Even more recent is the engineering that underlies ultra-processed food, ultra-processed entertainment, and ultra-processed connection. The evolutionary biologist Daniel Lieberman coined the term *dysevolution* for the incongruency between our hardwiring and our modern environments.

Once you are aware of dysevolution, you start to see it everywhere. It's why so many of us feel powerless in the face of short-term pleasures that our ancestors wouldn't have pondered in their wildest dreams, and also why we can't log off social media even though we feel terrible after extended periods of shitty flow. Fortunately, findings from the obscure field of *ecological psychology* can help us navigate the modern wilderness and its many traps.

Ecological psychology is the study of the intricate and often hidden relationships between ourselves and our environments. It suggests that the objects that surround us are not static; rather, they invite specific thoughts, feelings, and behaviors.

Experiments show that the mere sight of an object elicits brain activity associated with particular actions. For example, when we see a set of stairs, the networks in our brain responsible for coordinating the act of climbing begin to fire. When we see a bag of chips, we begin producing saliva. When we see a smartphone, we think about our email or who may have commented on our latest social media post.

The more we are around particular objects, the tighter the connection becomes. The first time a baby sees a chair, the motor programs in their brain do not automatically begin to fire in a sitting pattern. But by the time that baby is an adult and has sat in thousands of chairs, the sight of a chair invites a sitting response deep inside their brain.

This concept may seem esoteric, but its implications are practical: We should do what we can to surround ourselves with objects that invite our desired thoughts, feelings, and behaviors and eliminate those that do not. It may sound simplistic, but it makes a sizeable impact on the quality of our time and energy.

You could write out your values and goals and place them on your desk or mirror. You could also surround yourself with books, artwork, photos, or anything else that reminds you of your values and goals and that supports your path of excellence. For example, atop the desk at which I write are pictures of my children, memorabilia from my favorite authors, a stack of thought-provoking books, notecards containing inspiring ideas, and a sculpture symbolizing excellence and humanity crafted by an artist I admire. Above my desk is a banner that proclaims "Give a Damn." All of these objects prime me to do my best, most values-driven work. When I sit down to write, I am already a bit closer to where I want to be.

In his book *The Evolving Self*, the psychologist Mihaly Csikszentmihalyi, who coined the term *flow*, writes that the objects among which we work become "expansions of the self . . . things the mind can use to create harmony in experience."

Equally important is removing barriers and temptations.

Here are a few examples: When I read or write, I try not to bring my phone into the room with me. When I edit, I print out pages so as not to be distracted by anything on my computer. On Saturdays, a day I've set aside to connect with my family, my community, and myself, I place all my digital devices in a storage room in the basement. I call it a digital sabbath. These are just a few of many possibilities.

Before adopting these practices, I felt scattered, on edge, and psychologically exhausted more often than I would have liked. Not only did my ability to enjoy the present moment suffer, but my creative work did, too. It came to a head in 2022, when I published an essay—and really, a self-diagnosis—coining the term *internet brain*, in which I wrote, "Internet brain results from spending too much time online. It manifests as an inability to focus for long periods of time; a strong desire to 'check' something—be it social media, email, trending topics, or your favorite newspaper's landing page—even when you don't actually want to; a constant feeling of adrenaline that is somewhere between excitement and anxiety; a lack of patience for anything that is inherently slow; and a significantly harder time being present in offline life, such as constantly needing to pick up your phone." Writing that piece led me to a simple solution: I needed to mandate more time offline, something that would only occur with explicit constraints.

What makes sense for you will depend on where in your life you are pursuing excellence and what challenges you face. But it starts by acknowledging that it's impossible to overpower modern technologies with willpower alone. The aim is to design our environments as best we can, to create coherent ecosystems that support excellence in a fragmented world that often works against it.

Excellence as a Moral System—from Science to Philosophy

Think back to the neuroscientist Antonio Damasio's remark that even "in their un-minded existence, it turns out [bacteria] assume what can only be called a sort of *moral attitude*." He meant that bacteria innately move toward conditions that support their ability to stay alive and move away from conditions that degrade it. For all living organisms, from bacteria to human beings, every experience has an "inherent quality associated with it," explains Damasio. "It is the defining element of feeling. . . . It inevitably reveals the condition as good, bad, or somewhere in between. When we experience a condition that is conducive to the continuation of life, we describe it in positive terms and call it pleasant; when the condition is not conducive, we describe the experience in negative terms and talk of unpleasantness."

We've come a long way, including the development of minds that can imagine the future and goals that extend beyond mere survival. Today, most of us want to thrive, to move toward ever-increasing levels of homeostatic upregulation—toward fulfillment, resonance, satisfaction, and meaning; toward excellence.

Our biology and psychology are not neutral. They have an imperative to distance us from conditions that are detrimental to our flourishing (low quality) and draw us toward conditions that support it (high quality). This is every bit as true for our inherent drive to create good work and engage in worthwhile projects as it is for our avoiding famine or extreme climates. We know excellence when we see it, we are so deeply drawn to it, because excellence is in our nature. Herein lies the innate moral attitude of bacteria, and I believe our innate moral attitude as well.

In the next chapter, we'll explore this moral attitude in greater detail. We'll learn that long before we had the technologies and tools to develop a science of excellence, a few trailblazing philosophers were already imbuing it with significance, using many of the same terms that cutting-edge researchers do today. I'll argue that even beyond the boundaries of empirical science, striving for excellence is what we are here to do, an essential part of constructing a fulfilling and productive life.

Chapter Summary

- Much of what ails us—distraction, automation, mechanization, burnout, isolation, alienation, and emptiness—can be overcome by the pursuit of excellence.
- A common trap is when our short-term feelings are misguided and do not align with our values or long-term goals.
- Episodic future thinking (EFT) refers to the ability to imagine different futures and what they'll feel like. If we can connect our imagined futures to our values and goals, we give ourselves the best chance to stay on the path.
- Excellence may produce flow states, but it is a longer journey that requires sustained effort; it encompasses what happens before we get into the zone and what comes after, too.
- There are two kinds of burnout: conventional burnout, which comes from doing too much; and ***zombie burnout***, which comes from not doing enough of what lights us up.
- Effort pointed toward a goal we find meaningless will elicit an entirely different stress response than the same effort pointed toward a goal we find meaningful.
- We find meaning when our pursuits offer us autonomy, competence, and belonging and when they align with our values.
- The best way to overcome dysevolution is to design environments in ways that promote excellence—surrounding ourselves with objects that support our paths and removing barriers that detract from them.

Excellence is not a destination; it is a process of becoming.

The real reward isn't a bigger deadlift, a faster mile, or a sturdier table.

The real reward is that you become a better version of yourself.

3

The Philosophy of Excellence

In 1974, after being rejected by 121 publishing houses, Robert M. Pirsig's *Zen and the Art of Motorcycle Maintenance* was picked up by the small publisher William Morrow and Company. The book follows its author and his son on a motorcycle trip across the western states of America. It is part novel, part travelogue, and part philosophical treatise. At the heart of the book is *Quality,* which Pirsig defines as a pre-intellectual awareness between subject and object or actor and act.

When you value an activity, when you care deeply about it and give it your all, the space between you and whatever it is you are doing collapses, and Quality is high. When you don't value what you are doing, when you are distracted or going through the motions, there is a separateness between you and whatever it is you are doing, and Quality is low.

Quality captures the innate *rightness* we sense when we are absorbed in something we care about. It's also what attracts us to excellence in other people and their work. It's why we are moved by a stunning piece of art, beautiful architecture, or someone with master technique in their craft. In all of these instances, our awareness of Quality isn't something we think our way to. It is an embodied truth that we feel.

Seventeen years after Pirsig wrote *Zen and the Art of Motorcycle Maintenance,* he released his second book, *Lila.* In it, he continues to develop his theory of Quality, arguing that it is the driving force behind evolution. "Between subject and object lies the *value,*" he writes. It is "immediate" and "directly sensed. . . . If Quality were the primordial source of all our understanding then it followed that the place to get the best view of it would be at the beginning of history," he explains, before biologists came to the same realization. "Quality, selection, creates the world."

I first read Pirsig's work as an undergraduate student. It helped form my philosophy of life. He captured something essential about the experience of creativity, progress, and caring that no one had before. But his work was never taken too seriously by his contemporaries in academia, who dismissed it as being too pop, not sufficiently rigorous. However, we now know that his ideas track closely to the biological and psychological science that followed. All living species, including us, really are made to pursue Quality. The best view of it really does come at the beginning of history, when the first living organisms used it to sense and respond, to survive and thrive.

Quality is a driving force in personal and cultural evolution. It imbues our lives with meaning and significance. You sense it in the greatness of others, and you sense it when you're at your best. It's at the center of what scientists call homeostatic upregulation and what I call excellence.

In this chapter, we'll build on Pirsig's Quality to construct a broader philosophy of excellence that encompasses competence, connection, satisfaction, love, and other characteristics that are central to a good life. By the end, we'll be thoroughly prepared to move into the mindsets, habits, and practices of excellence that make up part 2 of this book.

Excellence as a Virtue

Pirsig wasn't the first to point toward excellence having moral significance. Over 2,500 years before, in ancient Greece, Plato and Aristotle wrote frequently about *arete*, which can be defined loosely as fulfilling one's purpose or function. In Greek philosophy, arete emerges when someone develops their innate faculties and skills and fully deploys them to pursue their values.

The term was used to represent both excellence and virtue, and these two meanings were often intertwined. The popular saying "excellence is a virtue," attributed to Aristotle, captures arete's double meaning. Excellence is not about perfectionism or obsessing over small details in every aspect of your life. It's about knowing your values and using your innate capabilities to pursue them. This, the Greeks believed, was central to a good life.

Around the same time the Greeks were focused on arete, philosophy was also burgeoning in Warring States China. There, disciples of Confucius emphasized *wu-wei* and *de*. Edward Slingerland, a professor of philosophy and Asian studies at the University of British Columbia, writes that wu-wei occurs when "a person is optimally active and effective . . . [in] a state of harmony that is both complex and holistic, involving as it does the integration of body, the emotions, and the mind." A person in wu-wei is also said to have de, which is translated as "virtue."

In his 2014 book *Trying Not to Try*, Slingerland notes that wu-wei is associated with "becoming part of something larger" or "fitting in with the universe." The resulting de represents "inherent goodness" or "the *right* way" to go about one's life. When someone possesses de, they are charismatic and attractive, as are their creations. Like Quality, you cannot easily define the experience of wu-wei or de, but you know it when you see it.

Ancient Greece and Warring States China are considered two of the most innovative and energetic philosophical epochs. In both traditions, the good life was not about attaining pleasure or avoiding pain. It was about striving for excellence: involved engagement in worthwhile pursuits that support your values and goals. Since then, countless philosophers have reached the same conclusion.

Baruch Spinoza, a seventeenth-century Dutch philosopher, developed an ethics around *conatus*, which is Latin for effort, striving, and impulse. For Spinoza, conatus symbolized an innate disposition to persist and enhance ourselves. In the eighteenth century, the German philosopher Immanuel Kant wrote extensively about *urteilskraft*, or an instinctive "judging capacity" that knows what is good without the need for logic or reasoning. Kant suggested that urteilskraft is what attracts us to beauty. He went as far as to say the feeling we get when we are creating or bearing witness to a beautiful masterpiece is evidence of God's existence.

In the nineteenth century, the legendary Russian author and philosopher Leo Tolstoy argued that "goodness is the eternal, the highest aim of life. However we may understand goodness, our life is nothing more but a striving towards the good. . . . Goodness is really the fundamental metaphysical perception which forms the essence of our consciousness: a perception not defined by reason." For Tolstoy, this deeply felt goodness is the connective tissue between an artist, their work, and everyone else who has an emotional experience receiving that work. The contemporary American philosopher Susan Wolf argues that an integral part of a meaningful life is "when one is actively (and lovingly) engaged in projects that give rise to a particularly rewarding type of subjective experience—it is, if you will, a high-quality pleasure."

Quality, goodness, urteilskraft, conatus, arete, wu-wei, and de are concepts that span time, geography, and philosophical traditions. Yet they all point toward excellence as a virtue, as an essential part of our nature, something we know deep down inside. Only recently has excellence fallen out of favor, the consequences of which are dire.

A Solution for Longing and Loneliness

During the process of writing this book, I had the experience of getting lost in thought while filling up my car at the gas station on an otherwise tumultuous day. It was a welcome moment of peace. Until, seemingly out of nowhere, I was interrupted by a celebrity, appearing on a high-definition screen on the pump, telling me that when I'm stressed I just need to repeat to myself, "Everything is figureoutable." That experience is emblematic of today's world, which cares not about our values, goals, or ability to think and live deeply but rather page views, screen time, and monetization. We are fed information—real, fake, somewhere in between—at such a dizzying pace it crowds out the time and space we need to think for ourselves.

Perhaps you are sitting down to work on something important to you, and your phone buzzes just as you are hitting stride. Or maybe you are sinking into a groove while listening to music, only to be interrupted by an advertisement for a mattress company. Individually, these intrusions are minor and harmless. But taken together, they have an overwhelmingly negative effect: They get between us and whatever it is we intend to do. In Pirsig's terms, they erode Quality. When these distractions become ubiquitous, they alienate us from our lives.

Research shows that alienation is associated with a range of negative consequences, including exhaustion, apathy, languishing, burnout, and even depression. But perhaps more than anything, alienation leads to existential loneliness, the feeling of losing one's center, of not being at home in oneself or the world. It's the opposite of the "situated" condition we explored earlier.

What do people do when they feel lonely and uncomfortable? Log on to the internet to buy stuff, post on social media, or doomscroll. Find their tribe and shout into the void. Consume substances that may bring relief in the short run but are detrimental in the long run. These activities offer pseudo-aliveness and pseudo-connection, short-term balms that exacerbate the underlying problem: a lack of intimacy, energy, and meaning in our lives.

This vicious cycle works perfectly for the merchants of the attention economy but terribly for everyone else. The greatest risk of modern overload is that we give up on excellence, Quality, and agency altogether. Instead, we go wherever the current takes us, like automatons floating along an algorithmic conveyor belt.

The only thing that separates us from this dystopia is ourselves.

That is, we have the choice of putting the phone in the other room, disconnecting from the internet, unplugging the cable, and so on. While there is nothing inherently wrong with occasionally scrolling social media or watching mindless television, the danger occurs when we aren't deliberate about, let alone aware of, how we are spending our time and energy. Excellence is impossible if we cannot distinguish between what is superfluous and significant, between what is trivial and meaningful. But if we can draw those distinctions and act on them, the rewards are great.

We don't grieve and we don't rest, we just choose the lie that feels the best.

These lyrics are from the opening track on the 2024 album *Visitor,* by the singer-songwriter John Moreland. The entire album is a gem. It wrestles with the overload, distraction, and remove so many of us feel from our lives.

When I spoke with Moreland about *Visitor,* he told me that he writes what he knows. Following the release of his 2022 album, *Birds in the Ceiling,* Moreland toured arduously and spent hours upon hours on the internet promoting his work. His attention grew more fractured as he became increasingly immersed in the outside world and its endless stream of judgments and opinions. As a result, he lost touch with himself. He began to feel alienated from his art and his life, which is to say he began to feel like crap. "I wasn't in a great place," he explained. "I felt far away from the music itself and the writing process, which are the things I actually love."

Moreland's rut reached its nadir in 2023 when he stopped touring and threw away his smartphone for *six months.* "I needed to not do anything for a while and just process," he reflected.

The result of his time unplugged was *Visitor,* perhaps his best and most intimate album yet. He wrote and recorded it in his home alone, with the exception of his wife and one other musician, each of whom sing on a single track. Moreland played all the instruments. The reason the album shines is because there is no space between Moreland and the work. No distractions. No extraneous bullshit. No noise.

"I'm proud of how it turned out," he told me. "The outcome was a reflection of the process that created it." Even the album's title, *Visitor*, reflects how Moreland was feeling when he wrote it—like someone who didn't belong in our chaotic world. But the process of making the album, of getting close and focused and working without distraction or alienation, allowed him to feel situated again, at home in himself and the world. It's what excellence is all about.

Moreland's experience isn't uncommon. Even if we aren't plugged in at the moment, spending too much time online puts us in a hypervigilant, attention-fractured state. It's a prime example of *internet brain*. We lose the ability to focus deeply on what is in front of us and fail to give our all to the things we care about most. This is a tragedy, full stop.

At the same time, we aren't going to rid ourselves of digital tools, and they aren't all bad either. After my third consecutive listen to *Visitor*, I messaged Moreland on the internet thanking him for his devotion to the craft and for writing such a profound album. He responded. It was a neat moment for both of us, and it led to a more authentic connection. Yet it's also true that if Moreland had not stepped away from all the noise that got between him and his work, there wouldn't have been such a poignant album over which to connect in the first place.

Pursuing the type of excellence that leads to a good, fulfilling life means identifying what matters to us and then deliberately eliminating distractions that get between ourselves and those pursuits.

Put differently, *excellence requires intimacy.*

Traditionally people think of intimacy in terms of being in a relationship with another person. While that certainly can be the case—and is a beautiful thing when it happens—you can also develop intimacy with an activity or craft. It is a sense of familiarity, respect, and attention that helps you feel connected to what you are doing and to yourself. It requires minimizing distractions and getting as close as possible to your pursuit. It's being in the pocket of a deadlift, song, or painting; it's being immersed in developing an idea, leading a team, or learning a new skill.

The intimacy of excellence is a powerful antidote to modern alienation and its associated feelings of disconnection, disassociation, and numbness. When you work toward intimacy with an activity or craft, when you throw yourself into something you care about and give it your all, you experience presence, depth, and aliveness. There's no way around it. Even if there were, it would defeat the purpose of striving for excellence to begin with: to create quality experiences for ourselves and for those who are touched by our work and our lives.

Satisfaction Through Competence

At one point or another, many of us have been frustrated by the out-of-touch actions or moral failures of prominent and conventionally "successful" people. I won't mention specific names, though I'm sure you could easily come up with your own. A while back, after one such letdown, I reached out to an older and wise mentor in search of solace.

Me: *I can't believe there are so many egotistical jerks. Why do all these people just completely lose touch? What is it about money or power or status that turns you into an asshole? Is it unavoidable?*

Mentor: *I am getting more weight equipment.*

We didn't talk much further on the topic. We didn't need to. I knew exactly what he was saying: Lifting weights offers a genuine source of satisfaction, so you don't need to chase the superficial variety.

"The only thing in life that's really worth having is good skill. Good skill is the greatest possession," Jerry Seinfeld once said in an interview with *The New Yorker*. "I know a lot of rich people. So do you. They don't feel good, as you think they should and would. They're miserable. Because, if they don't master a skill, life is unfulfilling."

Weight lifting and stand-up comedy are but two of many examples. Others include cycling, swimming, gardening, hiking, painting, climbing, sculpting, writing, music, and woodworking. What all of these activities share is a path toward satisfaction that is not wishy-washy or contrived. These pursuits do not entail grandiose visions of changing the world or reinventing industry. They are neither politically motivated nor do they require schemes. What they are is simple and real.

When the barbell drops, it drops. When you want to run a marathon in under three hours but go 3:04, the result hits you right in the face. When the legs of the chair don't fit into the base, the dilemma is as material as anything. The rhythm of the song cannot be faked. The blank page either fills with words or it does not. The tree either blooms or wilts. The joke either lands with raucous laughter or is met with awkward silence. It is hard to lose touch with reality or suffer narcissism when you are working with integrity on something concrete, when your successes are earned and your failures cannot be rationalized by technical jargon, corporate mumbo jumbo, or social media hot takes. Striving for excellence of this sort—doing real things, in the real world, with real results—keeps you grounded, both literally and figuratively.

In his 2009 book *Shop Class as Soulcraft*, the philosopher Matthew B. Crawford writes that "despite the proliferation of contrived metrics," many jobs suffer from "a lack of objective standards." Ask certain white-collar professionals what it means to do a good job at the office, and odds are they may need at least a few minutes to explain the answer, accounting for politics, the opinions of their boss, the mood of a client, the role of their teammates, and a variety of other external factors. But ask someone what it means to do a good job at their next marathon, on their next deadlift, or in their art studio, and the answer becomes much simpler.

"The satisfactions of manifesting oneself concretely in the world through manual competence has been known to make a man quiet and easy. They seem to relieve him of the felt need to offer chattering interpretations of himself to vindicate his worth. He can simply point: the building stands, the car now runs, the lights are on," writes Crawford.

Psychologists call this an *autotelic experience*. It represents the fulfillment and satisfaction that arise from doing a job well for its own sake—something that is only possible when there is a clear standard for what doing a job well means.

One of the humblest, most understated people I know is Blake, a technology leader at a world-class professional services firm.

During fire drills, when his colleagues are roiled in stress, he shows calm and equanimity. He doesn't let office politics keep him up at night. Whenever there is an ethical question, he does the right thing and makes it look easy. Blake is also a committed woodworker. It's not a coincidence.

When you are building tables in your basement, you are going to be humbled over and over again. Tables either wobble or they don't. You can't use power or money or relevance or fame or anger to make a shoddy table stand. Instead, you've got to put your ego aside, stay calm, adjust as you go, and work toward a solution.

There's a sign in Blake's shop that reads *Slow Down*. "The more you hurry the longer things take," he explains. "When I'm in the shop, I have to slow down and think several steps ahead. It makes me do a mental reset. It's the only time I go four or five hours without touching a device. It's a very grounding experience."

To be sure, there are still people who work toward concrete goals and are egotistical jerks nonetheless. But these people are rarely, if ever, going about it the right way. It's the athlete who cheats. The writer who plagiarizes. The artist who engages in fraud. When you go about your pursuits the right way—which, by definition, is what excellence is all about—you tend to gain at least some measure of morality. It's what each of the philosophers with whom we opened this chapter—albeit flawed humans themselves—were trying to explain in their own way.

"The place to improve the world is first in one's own heart and head and hands, and then to work outward from there," writes Pirsig in *Zen and the Art of Motorcycle Maintenance*. "Other people can talk about how to expand the destiny of mankind. I just want to talk about how to fix a motorcycle. I think what I have to say has more lasting value."

The Point of It All

What are we here to do?

Each and every one of us asks ourselves this question. I've made case that the answer is to strive for excellence, to figure out our values and goals and go after them with as much vigor and intimacy as we can.

Another fitting answer is that we are here to love.

But then again, what is excellence—what is caring deeply and paying attention and repeated practice and learning from failure and staying curious and steadfast dedication and getting really close and feeling deeply and showing up over and over again—if not love? And what is love if not a generative force that imbues our lives with Quality?

Perhaps excellence, Quality, and love are all one and the same.

Teaching. Coaching. Parenting. Leading. Doctoring. Nursing. Making art. Developing lifesaving medications. Inspiring through sport. Building bridges. Comforting a friend. Helping a neighbor in need. All of these are acts of creation, opportunities to pursue excellence.

Sometimes we leave behind work that is concrete: a book, a song, a world record, a piece of art, an invention. Other times we leave behind something less tangible: a story passed down for generations, a profound impact on someone in need.

Either way, when we pursue excellence, we transcend our small selves—the parts of us that worry, doubt, and fear—and enter into something larger, a dance with the universe. We experience the opposite of existential loneliness: We feel whole and settled, at home in ourselves and at home in the world.

In the preceding pages I've put forth a comprehensive theory of excellence: what it is, how it works, and why it matters. I've argued that we evolved for excellence, that from whatever angle we view it—biologically, psychologically, or philosophically—it is what we are here to do.

Now that we are equipped with a solid foundation, we'll move on to the essential mindsets, habits, and practices to cultivate excellence in our own lives.

Chapter Summary

- Multiple traditions spanning different geographies and thousands of years point toward excellence being a moral virtue that is key to a good life.
- When we aspire toward intimacy with what matters to us, we overcome alienation and disconnection and feel more at home in the world.
- Working on concrete tasks with objective and measurable results is a source of deep satisfaction and helps to keep us humble and grounded.
- The most raw and authentic pursuit of excellence begins to look a lot like love.

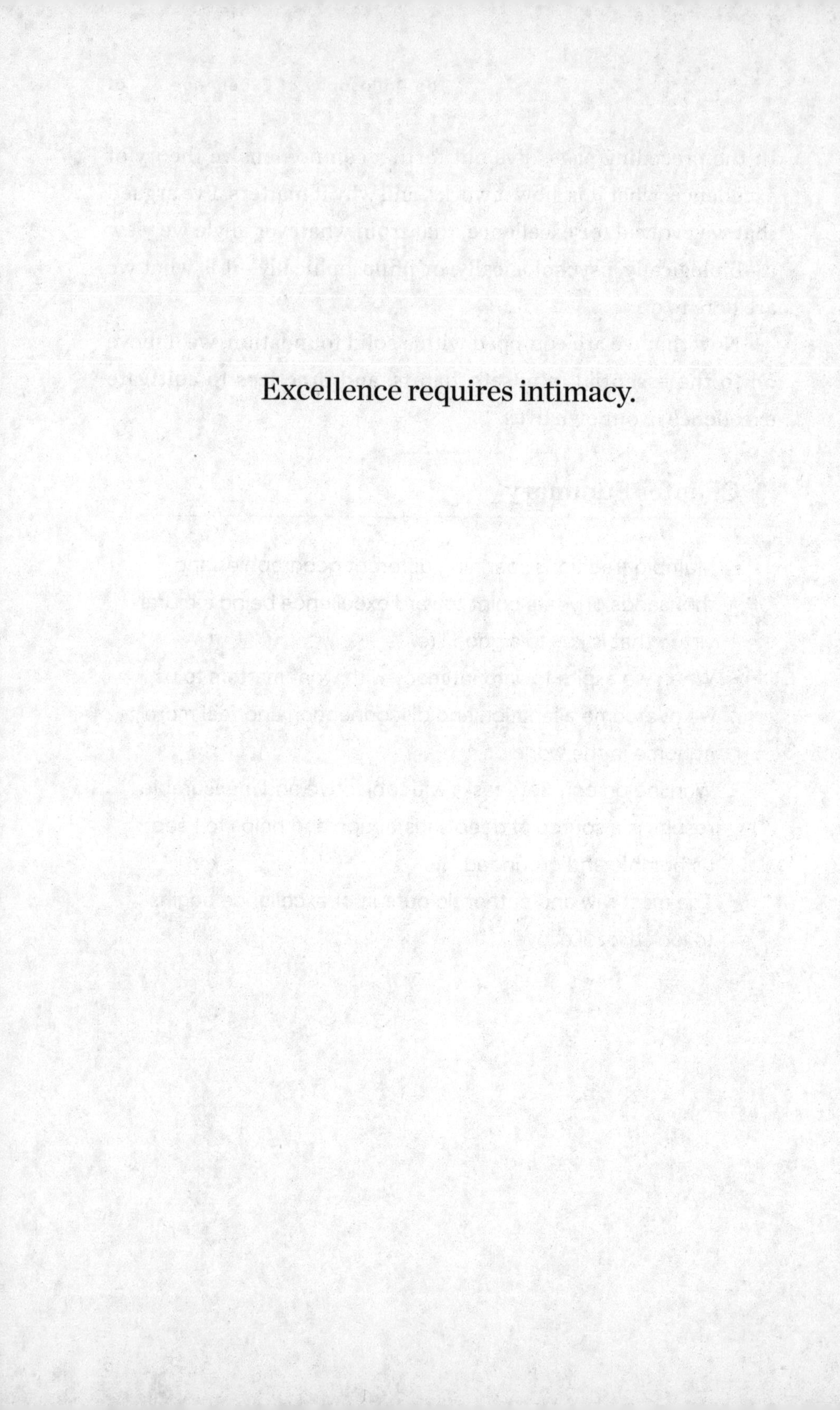

Excellence requires intimacy.

Part 2

MINDSETS, HABITS, AND PRACTICE

4

Care

You can possess all the knowledge and talent in the world, but it doesn't matter if you don't care. Caring drives everything, regardless of what it is you do.

Some days are better than others, but underneath it all, at the foundation of excellence, you'll always find a heartfelt concern, interest, and reverence for your pursuit. It's what allows you to give something your best effort, to stay on the path through ups and downs.

Caring is not always easy, and it doesn't happen overnight. In this chapter, we'll explore how to develop and sustain care. We'll also see how caring deeply and vulnerability go hand in hand. Just about anyone can put on an act of nonchalance or sit on the sidelines and pretend, but stepping into the arena and laying it on the line—pouring your heart and soul into something—requires strength and courage.

Cultivating Care—the Interplay of Fit and Grit

We place loads of pressure on ourselves to find the *one* thing. But when you talk to people who have achieved excellence, you almost always hear stories about experimenting with a variety of pursuits before narrowing in. Expecting to stumble upon an activity where everything clicks from the get-go is a surefire way to be disappointed.

Passion emerges over years, not seconds. It's rarely like lightning striking. More often it's like laying bricks for a house.

Lower the bar from perfection or unbridled enthusiasm and think about pursuing your interests instead. Do you see potential for growth in a certain field or activity? Does it pique your curiosity? Does it broadly align with your values? If so, see where it leads you. If not, don't be afraid to try something else.

The all-time great Roger Federer dabbled in soccer, basketball, skiing, wrestling, swimming, and skateboarding before concentrating on tennis. Amelia Earhart was a nursing aide and medical student prior to flying planes. Mega bestselling author Michael Lewis worked in finance for years before writing his first book.

Some world-class performers say, "I knew I was meant for this."

But a far more common story is: "I explored all sorts of activities and found my way to basketball or art or design or woodworking or whatever it may be. I kept at it because I enjoyed it and was pretty good, and ten years later, here I am."

Never give up and *Don't quit* are two of the most popular aphorisms in motivational speaking. Sometimes it's great advice, but other times it's harmful and gets in the way.

Consider the following examples: I struggled with math throughout my primary schooling. Because I tried exceedingly hard, I managed to get by. That is, until my senior year of high school, when I was placed in AP calculus. I lasted about two weeks before deciding to drop the class, lest I ruin my academic record with a failing grade.

In college, I thought I would major in economics and apply to the undergraduate business school. That plan came to an abrupt halt at the beginning of Intermediate Microeconomic Theory. I made it through four lectures before giving up on the course, along with any hope of a degree in economics or business.

Early in my career, I worked for a large consulting firm, and then at one of the most well-respected health-care systems in the world. The material was interesting, the money was good, and my colleagues were whip-smart. But even after a decade, I felt like I wasn't expressing my unique talents and potential. I couldn't get genuinely excited about a career in the field, so I stopped busting my ass in Microsoft Excel and PowerPoint for the next promotion and focused on developing my verbal skills instead.

As an athlete, I devoted ten years of my life to running. I trained hard and smart, and got pretty good. I loved the challenge and objectivity of the sport, and felt at home in the running community. But there was one significant issue: I kept getting injured. The more I improved, the more it felt like I was fighting against my body. I dealt with it for a while, including countless visits to doctors and physical therapists. But eventually, staying healthy became so tedious and challenging that I stopped running altogether.

In all three instances—education, career, and athletics—well-meaning people told me to tough it out. Winners don't quit. If only I would stick with it a little longer, they proclaimed.

But I knew myself well enough to realize I was no longer curious about what might happen next. I did not see a path toward growth. I had developed a strong intuition, based on plenty of evidence, that my mind and body were simply not meant for math, purely analytical thinking, or running, at least not at the level I wished to pursue them.

So I quit.

What happened next?

Academically I took a sharp turn toward so-called softer subjects. I found my way back to writing, which I'd enjoyed in high school. Professionally I started a blog and began pitching articles and essays to small magazines every week. I discovered I could write for hours on end and with great satisfaction. Athletically I began strength training. I learned that as much as my body rebelled against endurance sports, it welcomed being under a barbell and lifting heavy weights. Today, a few decades later, my primary craft (and livelihood) is writing, and the gym is one of my core communities and another avenue for mastery.

That is not to say these pursuits aren't without challenges.

Writing is hard nearly every time I do it. Getting to where I am now involved countless false starts, missteps, and rejections. Powerlifting contains plenty of frustrations and setbacks. Progress is always slower than I'd like.

There have been numerous opportunities for me to quit. But I can't imagine my life without writing or strength training. I enjoy them, I remain curious, and I still see potential for growth. As a result, I've come to *care* about them.

A study published in the journal *Nature Communications* set out to examine the career trajectories of artists, film directors, and scientists. Researchers found that hot streaks, "bursts of high-impact work clustered together in close succession," followed a predictable pattern. First, there is a period of exploration during which someone tries many different styles and methods. Then there is a period of exploitation, when a person settles in on a single approach and takes it all the way. Similar findings have been documented in athletes. On average, those who play multiple sports before specializing have greater longevity and achieve excellence at higher rates than those who focus on one sport from the beginning.

The route to deepening care (and to your best performances) often necessitates sampling widely and *exploring* until you find what is working—and then, once you've found that thing, narrowing in, cultivating it, and *exploiting* it. We excel at endeavors for which we are biologically and psychologically inclined. But without further development, innate talent only gets us so far. The age-old debate between nature and nurture would be better framed as *nurture your nature.*

Another way to think about cultivating care is the interplay of quit, fit, and grit.

Try a range of pursuits. Give it some time, but don't be scared to quit until you discover something where there is fit—an alignment with your values and natural abilities, a curiosity and potential for growth. Once you've determined fit, then the focus shifts to grit. Remember that passion and staying power are not automatic; you develop them by persevering through ups and downs.

If you over-index on grit before fit, you risk sticking with something for too long when quitting would have benefited you. If, however, you prematurely quit once you have fit—if you are always searching for perfection or telling yourself there must be something better around the corner—you'll never stick with anything long enough to excel.

Imagine if Federer hadn't quit soccer. Or if Earhart hadn't left medicine. But also imagine if Federer gave up on tennis when challenges arose. The same goes for Earhart and aviation. We would be missing one of the greatest athletes of all time and a trailblazing pilot.

Even when initial fit is there, it's just a seed. For an activity or pursuit to grow into something we truly care about, we've got to water it regularly with practice and attention. There are seeds of excellence in all of us. Our work is to discover which ones respond best to nourishment.

How do you know when you've found the right seeds?

You feel a pull to keep showing up. You may not get better every day, but there is a general sense of growth and progress. You may not love every moment, but there is an emerging satisfaction. Like a seed ready to sprout, you accumulate energy asking to be channeled. These are all signals to keep going. These are all signs of caring.

Overcoming Fear

Everyone remembers the kid in school who was too cool to try. Perhaps that kid was you. In reality, that kid lived in a combination of insecurity and fear. It was easier not to care than to care and risk failure. Many adults have yet to outgrow this tendency. Hence all the nonchalance and too-cool-to-care attitudes.

The biggest barrier to caring is fear.

Fear of failure. Fear of heartbreak. Fear of loss.

Things don't always go your way, and even when they do, they always change. The project you dedicated yourself to for over a year gets killed at the twenty-fifth hour, for reasons outside of your control. After months of preparation, blood, sweat, and tears, you lose the race. Your body ages, leaving no question that it's time to retire from the career that defined you. The movement to which you devoted your time, energy, and attention fails to accomplish its aim. The child you poured everything into raising moves out.

"Heartbreak is unpreventable," explains the poet David Whyte, "the natural outcome of caring for people and things over which we have no control. . . . Heartbreak is an indication of our sincerity: in a love relationship, in a life's work, in trying to learn a musical instrument, in an attempt to shape a better and more generous self."

A common defense is to move through life like the cool kid in school, to adopt an attitude of nonchalance, to prevent ourselves from caring by coasting or phoning it in, by keeping everything at arm's length instead of giving it our all. An attitude of nonchalance is akin to going through life with Bubble Wrap on. You make yourself safe, you protect yourself from scratches and bruises, from becoming worn in. But you never grow into who you really are, let alone express your true potential. The hurt isn't as bad this way, but neither are the joys. You end up missing out on the fullness of life. You sacrifice the pursuit of excellence—and the deep satisfaction it brings—for short-term safety and comfort instead.

If you want to have a rich and meaningful life then you have to expose yourself. You have to make yourself vulnerable. You have to care. There is no way around it.

The best athletes care deeply.
The best artists care deeply.
The best leaders care deeply.

The best coaches care deeply.
The best teachers care deeply.
The best healers care deeply.
The best parents care deeply.

You are not going to be the best anything with an attitude of nonchalance—including the best version of yourself.

How do you care deeply in a way that is sustainable? How do you make yourself vulnerable without it becoming overwhelming or burning yourself out?

You set boundaries. You maintain a sense of humor. You surround yourself with good art, good books, and most of all, good people. They remind you that caring is hard but worth it, and they offer crucial support on your path. You cultivate a sense of self that is larger than your ego, larger than the part of you who wants everything to go a certain way. You hold your care, ambition, and drive in a container of self-kindness, because if you cannot be kind to yourself when you suffer inevitable heartbreaks and failures, then you would never come back and risk putting it on the line again. You learn to tell yourself, *Caring deeply is hard, but we can do hard things.* This is what it looks like to have your own back, and it is only when you have your own back that you can truly become a badass.

You come to appreciate that you can either go through the motions and be superficially cool (but actually boring), or you can step into the arena, lay it on the line, and care deeply.

You come to appreciate that the things you care about are the things that break your heart are the things that give rise to excellence and fill your life with meaning and joy.

Identification and Identity

Caring deeply about something means beginning to identify with it. You start saying, *I am an artist* or *I am a chef* or *I am an athlete* or *I am a musician* or *I am a doctor* or *I am a parent*. It goes from being something you do to being part of who you are. This is to be expected. It is a sign of full engagement and intensifying care, a wonderful feeling. But it isn't without risk.

That's because when an activity becomes the *entirety* of who you are and something goes wrong, it upends your sense of self. Just knowing this could happen creates a source of unnecessary tension.

I've come to think about identity like a house: If you live in a house that only has one room, and it floods, then you have to move out of the house. It is a disorienting experience. But if you live in a house with multiple rooms, and one room floods, you can seek refuge in the other rooms while you repair the damage. The goal is to build an identity house with at least a few rooms, because you never know when one is going to flood and you'll need to find strength and stability in the others.

Example rooms include artist, athlete, bookworm, neighbor, leader, parent, musician, partner, chef, physician, nurse, teacher, writer, and so on. The rooms need not be the same size, and you don't need to spend the same amount of time in each. When you are pursuing excellence, you won't. The goal is *not* to be "balanced." You'll inevitably give your primary pursuit significantly more of your time, energy, and attention. You just want to make sure it's not the *only* room in your identity house. If it is, it makes you fragile.

It's good to be all in, you just don't want to be all in all the time.

The scientific term for this is *self-complexity*. It refers to diversifying the sources of meaning and identity in our lives. Studies show that when we increase self-complexity, we gain overall stability and are better able to weather challenges.

A common misconception is that by diversifying your identity you sacrifice in your primary pursuit. The opposite is true. Self-complexity allows you to identify with an activity without it becoming the *whole* of your identity. It ensures your care is not only strong but also enduring. It allows you to go all in, but with enough of a foundation that you can take risks. With self-complexity, you gain resilience, play to win instead of playing not to lose, and generally speaking, perform better. If you want to be really good at something, you have to be willing to fail. Being willing to fail is easier when you have a strong sense of self. Having a strong sense of self requires not fusing completely with a single activity or dimension.

If you realize you only have one room in your identity house, you can renovate and make additions. What interested you as a child? What activities do you wish you could do if you had just a little more time? What have you been setting aside that you know you probably shouldn't? These are all potential rooms in your identity house, outlets to keep you steady and strong.

In a future chapter, we'll discuss the paradox of going all in on an activity while also maintaining self-complexity. For now, it's important to internalize the broader notion of constructing an identity house that contains more than one room.

The pursuits we throw ourselves into and care about most will still break our hearts. That much is inevitable. But when they do, we'll have a better chance at putting our broken hearts back together and staying on the path. Knowing this will happen—and deciding to go forth anyway—is what having the courage to care is all about.

The Courage to Care

It's easy to sit on the sidelines. To think about the thing. Research the thing. Talk about the thing. Perhaps even dream about the thing. But these are all just ways of protecting yourself from actually doing the thing.

Another trap is doing the thing but not *really* doing the thing.

It's the person who has the website and clothes and haircut and notebook and pen and typewriter and all the other accessories they think a writer ought to have, but what they don't have is the guts to put themselves out there and write. They fear failure and rejection, or that their work won't be read widely. Spending all this time on the *act* of being a writer is what someone does when they are scared of *being* a writer. It's a protective mechanism, fancy and dressed-up Bubble Wrap.

Writing is an example I know well, but the pattern is universal. You see it in sports, cooking, music, photography, craftwork, and so on. You even see it in parenting. There is no shortage of elaborate ways to feign caring. Underneath each resides fear.

The word *courage* originates from the Latin root *cor,* which means "heart." It's fitting since excellence is not possible without caring, and caring is not possible without heart. Courage doesn't entail the absence of fear or doubt. It entails taking fear and doubt along for the ride and doing what we are meant to do no less.

When we cultivate a pursuit worth caring about, acknowledge and accept our vulnerability, and work toward identification without identity, we ready ourselves to step into the arena.

Chapter Summary

- It is impossible to be excellent without caring deeply about what you do.
- Passion doesn't strike like lightning; it develops over time.
- Think fit, then grit.
- Caring deeply makes you vulnerable; it's the price you pay for the meaning and texture it adds to your life.
- By building multiple rooms in your identity house, you become less fragile, gain resilience, and bolster yourself against setbacks.
- It takes courage to care.

The things you care about

are the things that break your heart

are the things that give rise to
excellence

and fill your life with meaning
and joy.

5

Goals

Every climber desires to reach the peak of a mountain, but every climber spends 99.99 percent of their time, energy, and attention on its sides. "To live only for some future goal is shallow," writes Robert M. Pirsig. "It's the sides of the mountain which sustain life, not the top. Here's where things grow. But of course, without the top you can't have any sides. It's the top that defines the sides."

Goals are like mountaintops. They are important insofar as they provide definition and direction for our journeys. They serve as targets, offering a wellspring of motivation. They keep us focused and prevent us from aimlessly wandering. Yet nearly all of our growth, development, and meaning occur not at the point of accomplishing a goal but during its pursuit.

Imagine that with artificial intelligence you could click a few buttons and, within seconds, "compose" an award-winning-caliber piece of music. Would this bring you fulfillment? It's doubtful.

Even the best outcomes are meaningless if we don't go through the process of creating them. It is *because* we've got to master the difficult and overcome challenges that pursuing excellence is so powerful and satisfying.

There is no greater illusion than thinking the accomplishment of some goal will change your life. What will change your life is how you are transformed in the process of going for it. When you select what goals to pursue, you are selecting what kind of person you want to become.

Goal Selection

Earlier I wrote about values, which are the guiding principles, the North Stars, for our lives. The ultimate aim is to live a life in accordance with our values, which means we want to align our goals with them as best we can.

The following are example values:*

Health: Taking care of my body and mind so I can be around and functioning for the people and activities I care about.
Craft: Pursuing excellence in a select few endeavors that matter most to me.
Family: Being there for my kids and partner.
Community: Showing up for the important people and places in my life.
Truth: Living with integrity and telling it how it is.

---

* Remember that if you are unsure of your values, there is an exercise in appendix 2 to help you determine them. See page 255.

Whenever we are contemplating a goal, the first thing we ought to consider is how it might impact our values. Will the goal allow us to practice our values? Will it hinder them? What trade-offs will the goal require we make? If we have no goals for the moment, we can consider what new ones might support the furtherance of our values.

Continuing with the examples above, a goal related to running may promote health, craft, and community. A goal related to writing might support craft and truth. Yet you would want to ensure that neither your running nor writing encroach on family or community involvement beyond what you are comfortable with. There are always trade-offs. The key is to be aware of them so you can evaluate and adjust as needed.

It's helpful to revisit your values at least once a year. Around New Year's Day is a good time for this. Research shows 91 percent of people fail resolutions, but taking an inventory of your values—and asking yourself if your goals support them—is almost always beneficial.

It's normal for values to evolve over time, but they certainly don't need to. Everyone's path is different. What matters is that instead of going on autopilot, you bring intention to this process. When you work on a big goal that aligns with your values, you are working on not only the goal but also yourself.

Striving Well

The best goals push us ever so slightly outside of our comfort zones. Too much of a challenge, and the result is anxiety. Too little of a challenge, and the result is boredom. What we want is for the challenge to match the outer edge of our skills. As our skills increase over time, we can ramp up the challenges. Think of it as a vector for growth.

A simple yet reliable way to enter this vector is to reflect upon where you are and where you want to be. Then ask yourself, *What is the next logical step?* The answer ought to provide a good starting point for homing in on an appropriate goal.

From there, a popular framework for setting goals is SMART. That is, goals should be Specific, Measurable, Achievable, Relevant, and Time-bound. A SMART goal essentially says, *I want to reach the peak of that particular mountain, which offers a suitable challenge and which I chose to climb for good reason, by the end of the year.* The peak could be a 405-pound deadlift, building a kitchen table, doubling the size of your business, or playing Beethoven's Piano Sonata No. 14.

SMART is a solid starting point, but as you'll see in the coming pages, it has limitations and only covers a fraction of what matters most for sustaining excellence.

When you set a big and meaningful goal, you must be wary of two traps.

The first is that you trick yourself into believing that just because you set a goal, something within you has changed. You tell everyone about your goal and think about it all the time. This brings excitement, but it doesn't mean you've accomplished anything of consequence. Cheap thrills come from talking and thinking about a goal. Lasting satisfaction comes from working toward it. Be careful not to confuse the two.

The second trap is that you become so fixated on achieving your goal that you rush the process, making poor decisions along the way. In mountaineering, this is known as "summit fever." Climbers who get overly attached to the idea of reaching a mountaintop in a specific timeframe often take unreasonable risks to get there. The consequences can be tragic. At worst, climbers get caught in bad weather and never make it down.

Summit fever isn't exclusive to alpinists. Research shows that athletes who are overly fixated on achieving their goals overtrain and get injured. Creatives who are obsessed with achieving their goals don't take appropriate time for rest, recovery, and renewal—the elements that can fuel breakthrough ideas. In more traditional workplaces, being obsessed with accomplishing goals often leads to burnout or worse. Researchers from Harvard, Northwestern, and the University of Pennsylvania found that overemphasizing goals, particularly those based on measurable outcomes, results in reduced motivation, irrational risk-taking, and unethical behavior.

Where does this leave us?

Instead of focusing on the goal itself, it's wise to break it down into smaller parts, making adjustments as we go. Doing so serves as a powerful focusing mechanism. It helps us stay present instead of worrying about what may or may not lie ahead.

Someone who understands the power of process better than most is Kaillie Humphries. Humphries is one of the winningest winter athletes ever, with three Olympic gold medals, one Olympic bronze, and five world championships to her name. She has dominated the sport of bobsled for nearly twenty years. When I spoke with her about her longevity in the sport, she told me that "success over two decades comes down to doing the right things every two hours. Olympic cycles are long. I do everything possible to break it down into manageable chunks. What do I need to do this year? What do I need to focus on this month? What is the aim for this week? What does that mean for this day? The bigger the goal, the smaller the steps," she explained.

Even within the confines of a single championship run—her sled making hairpin turns at upward of eighty miles per hour, the entire event lasting less than seventy seconds—Humphries is focused on the process. "I break the race down into segments that last a few seconds. This helps me stay in the present moment. Yes, it's one continuous run, but it's also a series of turns and corners and accelerations. Each one becomes its own challenge—*Here's the drive I'm going to do; here's how I'll approach corner one; in corner two, here's what I need to do.* My mindset for a single run is not really all that different than training for the Olympics as a whole. Everything is broken down into manageable parts. If I do what I need to do in each part, the rest takes care of itself," she says.

This mindset not only helped Humphries become one of the best athletes of her generation, but maybe even more so, it's helped in her personal life. It's an example of how excellence isn't just about what you accomplish but what you learn and who you become. Humphries's personal story includes overcoming countless challenges in her IVF journey. "In many ways, what it took to give birth to my son was harder than the Olympics—the emotional weight of it all and how badly I wanted it and how much was out of my control," she told me. "I kept trying to remind myself that all I could do was focus on the process. It was still hard in ways I can't even describe, but I think it helped."

We cannot control how someone receives our work, the weather on race day, the judge's mood during a competition, or all manner of other factors that impact outcomes, in work and life. Sometimes we do everything right and the outcome still doesn't go our way, even in spite of our deepest desires. All we can control is our process.

A process-over-outcomes mindset is essential for excellence. Here's how it works:

- Set a goal.
- Figure out the discrete steps to achieving that goal that are within your control.
- Mostly forget about the goal and focus on executing the steps instead. Judge yourself based on the level of presence and effort you are exerting in each moment.
- If you catch yourself obsessing about the goal, use that as a cue to ask yourself what you could be doing *right now* to help you move forward. Sometimes the answer may be nothing at all. In that case, rest.
- Whether you succeed or fail, learn from the outcome, refine your process if necessary, and move forward.

Specific, measurable, and time-bound goals are most powerful when you are relatively new to an activity. However, as you make progress, the best goals tend to be broader and more open-ended.

For example, a novice at powerlifting may benefit greatly from the specificity and structure of a goal like *deadlifting 405 pounds by the end of the year*. But for someone who has been competing for more than a decade and who is pushing the limits of their potential, a specific, measurable, and time-bound goal may be too narrow and constraining. A better goal might be more along the lines of *get as strong as possible*.

After crossing certain performance thresholds, it becomes harder to improve. The next incremental gain is always more challenging than the last, and progress is less predictable. The deeper we go in our respective crafts, the less we require the rigidity and accountability of strict and specific goals. Instead, we benefit from the freedom to refine and adapt.

There are periods on everyone's path during which SMART goals are hugely beneficial. And there are periods on everyone's path during which SMART goals get in the way. It all depends on where you are. What remains crucial at all junctures is a solid process. This is true whether you are an enthusiastic beginner or a world-class performer for whom visible improvement is slow and minuscule.

The people who achieve big and audacious goals are rarely obsessed with achieving big and audacious goals. They are focused on the path, on the process. They weather ups and downs. They recognize meaningful progress does not come from intensity or heroic efforts on any given day but from consistency and discipline over months and years. They understand it is important to pick the right peaks but even more important to climb the right way.

Dig Where Your Feet Are

For as long as he can remember, the basketball superstar Ray Allen dreamed of winning an NBA title. He was the consummate professional, never missing a practice. He spent hours upon hours refining his already world-class jump shot. He made all-star games, won three-point shooting titles, and even starred in a popular Hollywood movie, *He Got Game*. Only one accolade evaded him, and that was winning a championship.

So when he finally won one in 2008, after twelve years in the league, you'd think he would have been elated. And briefly he was. But it didn't take more than a few days for that feeling to change.

"I felt empty," Allen recalls. "It had to do with having always believed that when you win a championship, you're transported to some new, exalted place. What I realized was that you are the same person you were before, and that if you are not content with who you are, a championship, or any accomplishment, isn't going to change that."

Allen learned the hard way a universal truth: Fulfillment, happiness, and satisfaction are not attributes you gain from achieving a goal; they are states that arise in the process of going for one. This remains true whether you are trying to win an NBA championship, open a restaurant, attain a promotion, hit a bestseller list, or achieve an Olympic medal.

If you develop a mindset *If I just do this, or just accomplish that,* THEN *I'll arrive,* you are in for a rude awakening. We never arrive. The goalpost is always ten yards down the field. The human brain did not evolve to be satisfied. It evolved to strive.

Psychologists call it *the arrival fallacy*. "We live under the illusion—well, the false hope—that once we make it, then we'll be happy," says Tal Ben-Shahar, the psychologist who coined the term. However, the work of Shahar and his colleagues shows that when we do make it, when we achieve our big goals and finally arrive, we may feel a temporary increase in happiness but that feeling doesn't last. If we expect external achievements to provide lasting fulfillment, we'll be surprised at how hollow we feel when we achieve them.

To be clear, it is normal to desire great outcomes. In many cases, they are legitimately important. Winning awards, hitting growth targets, publishing your work, and earning promotions may provide financial security or open the door to further opportunities. But what great outcomes won't do is provide lasting fulfillment—that doesn't come from wanting something and then getting it. It comes from digging where our feet are and finding growth and meaning in the process.

"I thought success would feel a certain way, but it has been a different experience than what I had imagined," the violinist Hilary Hahn told me of winning her Grammys. "It's helped my career in many ways, to be sure—but it hasn't fundamentally changed who I am or how I want to express myself. That is something I'll always have to figure out for myself, and it will always have to come from within."

This is not inspirational fluff. I've spoken to hundreds of world-class performers. When they reflect on their most memorable moments or stretches, they don't spend much time reminiscing about what happened at the peak. What they remember and relish most is what happened on the climb.

Best at Getting Better

The ultimate way to embrace a process-over-outcomes mindset is to stop worrying about being the best and instead focus on being the best at getting better. Being the best is a flash in time. You get there or you don't. Either way, it comes and then it goes. But being the best at getting better—that's a commitment to mastery that lasts a lifetime.

"Better" is not just about how fast you can run six miles, how many deals you can close per week, or how many articles you publish in a year. While those sorts of objective accomplishments contribute to "better," they comprise only a part of it.

Better is also about becoming stronger, kinder, and wiser.

When I think about my life, it was during the seasons when my objective performance was the worst that I ended up growing the most. During injuries and failures, both physical and mental, I felt awful. I wasn't more productive or a higher performer. But looking back, it was from the lessons learned during those stretches that I truly became better. Not a better writer or a better athlete, but a better person.

Deals closed, awards won, and promotions earned represent only a small portion of the balance sheet that is one's life. So yes, set meaningful goals and pursue them with everything you have. Doing so is core to excellence. But keep in mind that the real reward isn't just what happens when you reach the top of the mountain. It's who you become on the sides.

Chapter Summary

- Goals serve as targets and offer a wellspring of motivation.
- It is important that your goals align with your values.
- The best goals push you ever so slightly outside of your comfort zone.

- SMART goals are great—until they become too narrow or constraining.
- We never arrive; expecting to is a trap that leads to perpetual disappointment and emptiness.
- Growth, development, and meaning occur not at the point of accomplishing a goal but during its pursuit.
- Adopt a process-over-outcomes mindset.
- Be the best at getting better—and remember that better means stronger, kinder, and wiser, too.

There is no greater illusion than thinking the accomplishment of some goal will change your life. What will change your life is who you become in the process of going for it.

6

Consistency

Everyone loves to broadcast their heroic efforts. The string of all-nighters prior to a product launch. The workout ending in vomit. The all-consuming creative burst that lasted an entire week. No doubt, there is a time and place for these feats—but they ought to be a rarity. The toll they exert on you is great, and thus they are not sustainable.

One area where this is readily apparent is sports. Research shows injuries are most likely to occur when an athlete increases their training load too quickly. The best way to make progress and avoid injuries is to slowly raise the intensity and duration of training over time. In one study, when acute workload (the combination of the intensity and duration an athlete exerts over the course of a week) was more than twice as much as chronic workload (the average intensity and duration an athlete exerted over the prior four weeks), they were significantly more likely to get hurt.

Though the exact sweet spot for accumulating workload varies based on the athlete and their sport, the idea is that you don't want to increase any given day's exertion too much beyond the average of the past month's. Athletes who sustain massive workloads didn't get there overnight. They got there by being consistent over months, years, and in some cases, decades.

This principle applies beyond the playing field.

The way to build anything worthwhile is gradually, not erratically.

It doesn't matter if you are a creative, entrepreneur, artist, scientist, or student. People love the acute thrill of pushing over the edge, but they neglect its long-term impact. Whenever this happens, burnout, injury, or illness looms around the corner. The saying "Go big or go home" has a nice motivational ring to it, but the truth is those who go too big, too often, almost always end up home. A better guiding standard: Small steps taken regularly lead to big gains.

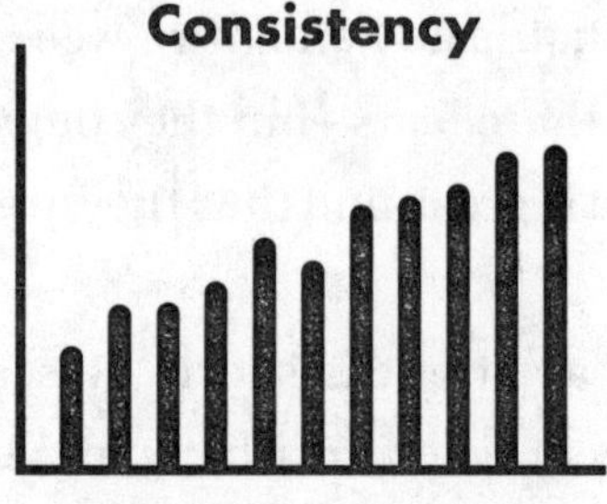

Repeated small efforts compounded for big gains.

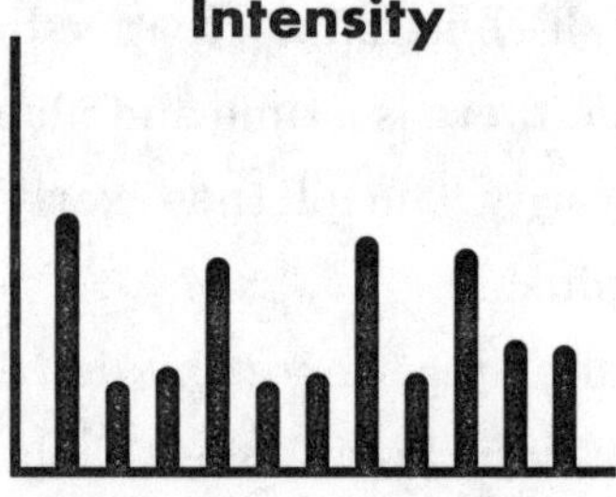

Heroic efforts followed by injury or burnout repeated.

Performative greatness is obsessed with heroic days. Actual greatness concerns itself with heroic decades. The goal is not for any single outing to be extreme but for the totality of effort over years to be extreme.

Excellence does not come from intensity.

It comes from consistency.

Little by Little Becomes a Lot

Long before he experienced success and amassed staggering wealth, a young Warren Buffett would tell his family and friends, "Do I really want to spend $300,000 on this haircut?" The haircut he was referring to didn't actually cost six figures. It cost more like five dollars. Buffett's point was that if he cut his own hair, he could save and invest the monthly five-dollar expense. Over the course of his life, all of those seemingly insignificant five-dollar investments would have netted him thousands.

Though Buffett's example may be unusual, he was speaking to the law of compounding: Small investments made regularly over time build upon themselves and grow into something big. Compounding doesn't just apply to excellence in investing. It applies to excellence in anything.

Every day we show up and work on our craft we are making a deposit in the bank. This doesn't mean that progress is always linear. We'll have good days and bad days, exciting days and boring days, and everything-in-between days. Every one of these days counts. What matters is continuing to make the deposits.

"Athletes feel the need to go search for confidence, and they do this by absolutely crushing themselves—but it takes so much time to recover from these sessions, or, even worse, they get injured," the elite triathlon coach Matt Dixon told me. "But when an athlete believes in overall progression, they don't need to seek validation in mega-workouts. Instead, they execute within their capacity for weeks, months, and even years."

It's not just athletes. It's all of us.

It's the writer who has the confidence to stop one sentence short, the runner who has the confidence to stop one rep short, or the entrepreneur who has the confidence to stop working thirty minutes before exceeding their limit. Staying consistent often requires demonstrating a bit of restraint today so that you can pick up where you left off tomorrow.

Regression to the mean describes the short-term tendency of a dynamic system to return to its average state. After a string of superb days, we tend to drift back toward our baseline performance. After a string of bad days, our performance tends to improve. Regression to the mean says that we should not worry so much about outlier efforts and instead focus on becoming a better average over time, prioritizing progress over perfection.

Becoming a Better Average

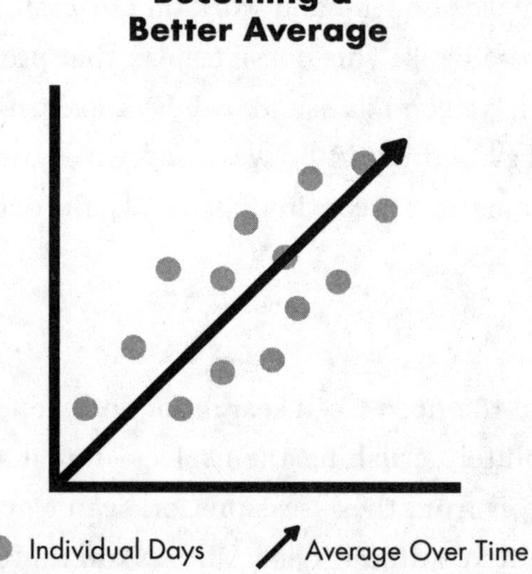

Unfortunately, so much of what we see publicly promotes the urge to shortcut a consistent approach. It's never been easier to seek and receive external validation for a huge effort, and it's never been easier to compare what we are doing to others. A prime offender is social media, where everyone is keen to post the greatest highlights of their day—the 4:00 a.m. wake-up, the hype speech about working "hardcore." What we don't see is how they felt the following day, let alone all the airbrushing that went into the original image or video.

Excellence demands a shift in attitude.

We need to worry less about swinging for home runs and more about repeatedly putting the ball in play. We must defy the urge to kill ourselves so we can have a few notable days and instead think about creating a notable body of work.

Become known for your consistency.

Raising the Floor

The French chess grand master Maxime Vachier-Lagrave, or MVL as he is widely known to his fans, holds the seventh highest rating of any chess player in history. He's most renowned for his tournament play. Tournaments involve multiple games in a single day and can last up to a week. When I spoke with MVL about his mindset, he shared how much he loves the energy and thrill of competition. Whereas some struggle in high-stakes moments, he thrives in them. He also shared a key to his success: It's about not just what happens when he's at the top of his game but also what happens when he makes an error—or in chess terms, when he blunders. "Of course I make poor moves from time to time. It's part of the game," he explained. "You've got to recognize the blunder, learn from it if you can, and then shift your focus to the present moment. You don't want to get into a mindset where you are playing past moves. You need to be able to quickly move forward. You need to play the board in front of you."

It is easy to focus when you are motivated, at your best, and everything is clicking. It is harder to focus when you are feeling off and as if you are moving against the current. But what you do on your not-so-great days may be most critical of all. Nothing brings down an average like a zero—it's basic math. So you want to avoid zeros whenever you can. Herein lies an idea called *raising the floor*, which entails making your bad days just a little better.

When I'm working on a book project, I try to write at least one thousand words on specified days. But sometimes it's just not there. Rather than phone it in and take a zero, I acknowledge I'm having a bad day, release from the original target, and attempt to get two or three hundred words out instead. It's no different from an athlete who is having an off day at the track adjusting their running pace, or MVL's commitment to playing the board in front of him. It's an approach that is applicable to anything. Raising the floor means shifting our expectations to account for what we've got to give in the moment and then doing our best to give it—remembering that even the smallest efforts are deposits in the bank, and they all drive compounding in the future.

A significant part of what separates great performers from everyone else is how they respond to micro-adversities. It's when a leader giving a presentation fails to deliver the message as she had hoped, when an athlete runs 15 percent slower than his planned target during a workout, when an artist throws out nearly all of her output from a given day, when a physician misses a diagnosis, or when a craftsperson realizes after six hours of manual labor that he has to dissemble the desk and begin again. These situations happen to everyone; they are unavoidable, even for the best.

What comes next is key.

We can either freak out, catastrophize, and obsess over the subpar effort or we can accept that it happened, learn from it if possible, and also realize that sometimes there is no silver lining and we just have to move on. It's the difference between having a bad moment, having a bad day, and having a bad day turn into a bad week.

Research shows that when we catastrophize and view missteps or bad days as threats to our overall progress, our bodies release the stress hormone cortisol, which is associated with chronic inflammation, anxiety, and decreased future performance. But the same research shows that if we can view missteps or bad days as minor speed bumps and see them as challenges to overcome, then our bodies release the hormone dehydroepiandrosterone (DHEA), which has been linked to a reduced risk of chronic inflammation, decreased anxiety, and improved future performance.

Short-circuiting the negative effects of missteps and bad days relies on a mindset shift that prioritizes consistency over intensity. When we embrace this mindset, we acknowledge that not every rep or every day is going to be great, and that's okay. We become less concerned with each moment's or each day's measurable performance and more concerned with the effort we gave. We realize that if we can eke out even a bit of good work on a bad day, then we're making progress. Even if we can't eke out any good work, but we don't let a bad day devastate us and ruin our week—then that, too, is making progress. "It's one thing to play well in a single game or tournament. It's another thing to play well for years and years," says MVL. "The latter requires a commitment to continuously showing up and playing what is on the board."

Rory McIlroy entered the final day of the 2025 Masters golf tournament in contention for the win. If he could pull it off, it would mean a career grand slam, placing him on Mount Rushmore of the sport's all-time greats. He'd been trying for over a decade and had fallen short in a variety of ways, but today felt different. He was in the lead, and playing strong.

That is, until hole 13.

McIlroy began the hole up three strokes, but then made a terrible double bogey, sacrificing his lead. The shot he missed may have been the worst of his career. He hit a straightforward approach into the water. It was an awful ball—a devastating unforced error. The crowd fell silent. The television commentators shouted "Oh no!" and were left speechless. It seemed as if he was going to fall apart. But then, on holes 15 and 17, McIlroy made what may have been the two best shots of his career, placing him back in the lead. I distinctly remember watching the drama unfold with my family from the hotel at Orlando International Airport during a one-night layover. After the seventeenth hole, my then-seven-year-old son, enthralled by the emotional roller coaster, asked, *Dad, is golf always like this?* To which I responded, *No, never*. My wife turned to me and laughed, and at the same time, we both said, *I think McIlroy's finally got it.*

On the eighteenth and final hole, he had an "easy" putt—though any golfer will tell you that, in reality, no such thing exists—to win. He missed. He choked, the consequence of which was that he now had to compete in a sudden death playoff. After all that he had been through on this course over the last five hours, over the last *ten years*, he was just five feet away from the win. Still, he fell short.

And yet throughout it all, he held it together. Maybe just barely, but together. No thrown clubs. No head hanging low. No deciding to step off the psychological roller coaster and quit. And then, in the first hole of the playoff, McIlroy played textbook golf, and he won the Masters.

Golf may be the ultimate test of staying consistent, responding to micro-adversities, and regulating your emotions. Over the course of a tournament, which spans four days and seventy-two holes, nothing ever goes to plan. You prepare. You practice. You visualize. And then stuff happens. The real game isn't the one you'd hoped or wished for, it's the one in front of you. And the difference between those who collapse and those who rise? It's how they respond, especially when things don't go their way.

What's true in golf is true in life.

Each and every one of us will face setbacks, unforced errors, and moments when our emotions flare and our plans fall apart. What matters most is how we respond—again and again and again. It's called having a *next-play mentality*, and it's a central feature of consistency.

A next-play mentality says that when things are going great, you should ride the momentum, but at the same time not get complacent or stuck in the prior moment. When things are falling apart—when you make an error, when you find yourself losing—learn from what happened if you can, but then forget about it and move forward. Throughout it all, keep showing up as best you can.

When our North Star is consistency, we adopt a short memory, which ironically is key to playing the long game. We forget about what happened on a great shot or day and get back to work. We forget about what happened on a bad shot or day and get back to work.

It's reminiscent of a popular Zen koan:

Before enlightenment,
chop wood, carry water.
After enlightenment,
chop wood, carry water.

The Usefulness, and Myth, of "1 Percent Better Every Day"

"Get just 1 percent better every day" has become a popular saying in the canon of self-improvement. On its face, it points toward much of what we've discussed above. We need not crush ourselves; we just need to show up and get a little better every day. All of those 1 percent gains compound, resulting in something massive. The only problem is that when taken too literally, eventually this slogan backfires.

Try telling MVL, Rory McIlroy, or someone who composes great music, writes award-winning essays, builds impeccable code, runs a 2:10 marathon, or is chef de cuisine at a Michelin-star restaurant that they should get 1 percent better every day. It's not going to happen. Holding themselves to this standard would be forever frustrating.

A more accurate description of progress looks like this: When you are new to an activity, you might get 100 percent better every day. As your skill increases, the gains become more incremental—10 percent, 5 percent, 1 percent, half a percent, a quarter of a percent, and so on. Eventually, improvement becomes so small you can't even perceive it. At this point, you might find yourself on a plateau for a few days, weeks, or maybe even months. Then suddenly, perhaps when you are least expecting it, you experience a breakthrough.

In other words, progress is nonlinear.

The implication of this truth is both simple and profound. If you become addicted to observable progress, if you require immediate and visible gains to fuel your consistency, then you will not last very long in whatever it is you do. It is easy to stay consistent when you are getting 1 percent better every day. It is harder to stay consistent when progress is less apparent. It's the primary reason so many people flame out after the honeymoon phase of a new activity. What separates the wheat from the chaff is what happens when you *stop* getting 1 percent better every day.

"Early in my career, when I was on the way up, the competition may not have been as strong, and I was winning all the time," MVL told me. "But now that I'm at the top, it's all very hard. I'm always playing against the best in the world. My longevity in chess has less to do with how much I love winning and more to do with how much I love competing. You've got to have that internal drive to be the best, and you've got to have a passion for what you're doing."

The most realistic and effective approach for sustaining excellence involves viewing the work as an ongoing practice, measuring and judging your level of attention and effort, and letting progress be a by-product of that. Focus less on any single result and more on the trend line.

As your skill rises, it becomes increasingly important to get comfortable with plateaus. McIlroy plateaued for eleven years before winning the Masters. But throughout it all, he kept showing up. "When you have a long career like I have had, luckily, you sort of just learn to roll with the punches, the good times, the bad times, knowing that if you do the right work and you practice the right way, that those disappointments will turn into good times again pretty soon," he says.

Instead of relying solely on visible progress for motivation, you've got to find joy in the work itself and the community in which you do it. You may not always *see* progress to indicate a good day, but you gain a *feel* for it. Were you focused? Did you give what you had to give on the day? Did you work hard but not overshoot the target? Can you come back tomorrow ready to go? This internal gauge becomes the most vital of all.

"One percent better every day" works as a mantra for consistency early on, but eventually you'll want to release from it—perhaps in favor of something like "I show up consistently and give what I've got."

Keep Pounding the Stone

Imagine there is a giant stone and your goal is to break it.

You start pounding with a hammer.

The outer layers crack easily. You are making progress. But as you get deeper, the sediment thickens. There are strings of days when it feels as if your hammering is productive, yet the stone does not give at all. There are also days when you are sick or extenuating circumstances get in the way of your pounding. Instead of freaking out, you adjust your plan and get on with the show. You take time for rest and renewal, too. All of this is part of your program. What allows you to be so rugged in your consistency over the long haul is that you are flexible enough day-to-day.

You keep showing up and pounding the stone until on one fateful day, after months or maybe even years of effort, the last layer finally cracks.

You celebrate.

And then you move on to the next stone.

The stone is writing a book. Making an Olympic team. Opening a restaurant. Starting a business. Winning an important case. Winning the Masters. Finishing your medical residency. Making a groundbreaking scientific discovery. The next stone represents the next challenge on your path of excellence.

Throughout it all, consistency is key. When you win at consistency you give yourself a good chance of winning at everything else.

What is winning at consistency?

It is showing up. It is getting started. It is giving what you've got to give on the day. It is playing the board in front of you. It is coming back tomorrow. It is doing this over and over again.

Chapter Summary

- Worry less about heroic days and more about creating a heroic body of work.
- Nobody can be consistently great, but you can become great at consistency.
- Use the law of compounding: Small steps taken regularly result in big gains.
- Raise the floor. Remember that what you do on your bad days may be more important than what you do on your good ones.
- Beware of addiction to observable progress. You get 1 percent better every day until you don't—what happens next is key.
- The more skilled you become, the more important it is to frame the work as an ongoing practice, measure and judge your level of attention and effort, and let progress be a by-product of that.
- Show up every day and give what you've got.
- Become known for your consistency.

The secret is there is no secret:

Consistency over intensity.

Fundamentals over fads.

Progress over perfection.

Over and over again.

7

Trade-Offs

Nearly all the great performers I've gotten to know, from athletes to artists to computer programmers to entrepreneurs, report a direct line between being happy and fulfilled, and going all in on something they care about. Consider Chelsea Sodaro, a world-champion triathlete: "The route to fulfillment is to find the things that light you up and pour your all into them—you've got to keep the main things the main things," she told me. Michael Joyner, a distinguished physician and researcher at Mayo Clinic, explained, "You need to be a minimalist to be a maximalist. If you want to be really good, to master and thoroughly enjoy one thing, you need to say no to many others."

Conventional wisdom says we should strive for balance. The term has become a staple of the self-help industrial complex. It graces the cover of countless books and magazines. Yet over the years, I've noticed something interesting: The times when people say they are at their best and feel most alive are also times when they are the *least* balanced.

Falling in love. Starting a company. Writing a book. Opening a restaurant. Training to set a personal record in a triathlon. During these bouts of full-on living, we are consumed by our activities. Trying to balance and devote equal amounts of time and energy to other areas of our lives would feel forced and detract from the formative experience.

Excellence requires reimagining how you think about balance, if not discarding the term altogether.

Beyond Balance

It is impossible to be the perfect partner, friend, parent, athlete, and employee; to read fifty books every year; to keep a clean house; to stay up to date on pop culture; to do all the self-care routines, and on and on and on. And yet this is what balance, at least as it is popularly conceived, has come to represent. By trying to be balanced, we end up driving ourselves crazy.

For most people, it is better to do a few things well, with full attention and care, than it is to spread yourself thin and do a bunch of things with average attention, energy, and results.

"It hasn't been easy, but over time I've learned what matters to me, and I've also learned how to be okay saying no to pretty much everything else," Sodaro told me. "Whereas I used to have fear of missing out, now I am comfortable in my own skin. I can own that my family and triathlon get nearly all of my time and energy at this stage of my life, and that's okay. It won't be like this forever, but for now, this is what it takes to pursue the sport I love at the highest level and also be the kind of parent to my daughter that I wish to be. Some people might judge me or think my life is boring, but that's fine—I'm okay with it," she explained.

Being a mature adult means acknowledging that trade-offs exist and being willing to make them. It is a wonderful feeling when you give yourself permission to devote your time and energy to what matters most—even, and perhaps especially, if that means de-emphasizing other pursuits or maybe even leaving them behind altogether. The key is that you don't go about it mindlessly.

Two concepts can help. The first is establishing a minimum effective dose: the amount of time necessary to stay connected to each of the central rooms in your identity house. It's okay to spend a disproportionate amount of time in one room, so long as you don't let others get moldy. It's also okay to let go of certain activities, so long as they aren't integral to who you are or wish to become.

Imagine you are going all in on a big creative project. It is your foremost priority. Needless to say, you'll be devoting most of your waking hours to the creativity room of your identity house. Even so, you may determine that no matter what, you'll still exercise for at least thirty minutes every day, be with your family from 6:30 to 8:30 at least four evenings per week, and make plans with friends no less than twice a month. These represent your minimum effective doses for health, family, and community.

When we are wholly immersed in a particular activity—whether it's sport, art, business, or something else—it's all too easy to let the inertia of the experience carry us forward without stopping to evaluate what we're sacrificing along the way. It's how people destroy their health, lose their relationships, and burn out.

Instituting minimum effective doses ensures that we don't enter a deep rut or make irreversible mistakes. It helps us to stay in touch just enough with other parts of ourselves so that when the time comes, we can find our way back to them. It ensures that we don't neglect foundational habits and practices, like exercise and social interactions. In the moment, these habits may seem superfluous and as though they are detracting from our main focus, but over time, they are what make our efforts sustainable. Excellence requires sacrifice and incredibly hard work. But it also requires taking care of ourselves.

Clear minimum effective doses also empower us to give our primary pursuits our full attention when we are working on them. That's because we don't have to waste cognitive or emotional energy worrying about letting other things fall through the cracks. So long as we are meeting our minimum effective doses, anything goes. When we are *in* on our main pursuit, we can really be *in*.

The second concept essential to prioritization is what psychologists call *internal self-awareness*, or the ability to see ourselves clearly by assessing, monitoring, and proactively managing our values, emotions, and behaviors. It's about creating the time and space to know ourselves, check in with ourselves, and then prioritize how we spend our time and energy accordingly.

Practicing internal self-awareness allows us to evaluate and reevaluate the trade-offs inherent to pursuing excellence. It ensures that we are making conscious decisions about where our priorities lie and thus decreases the chance that we'll have regrets about what we did and did not do. It helps us realize when our identity may be getting too interwoven with a specific activity, and that in some instances—writing a book, the first few months with a newborn baby, completing a PhD, or trying to make an Olympic team, for example—our lack of balance may feel excessive, but that's okay because it is intentional and values-driven.

Internal self-awareness does not come easily. Paradoxically, some of the best ways to access it involve mentally stepping outside of our current "self." Psychologists call this *self-distancing*. Pretend a friend came to you for advice. They are struggling with wanting to go all in on an activity but concerned about leaving other things behind. What would you tell them?

Or imagine an older and wiser version of yourself looking back on your current self. What might they think about your priorities and how you are spending your time? What counsel might they offer? Finally, you could reflect on your own mortality, which has a piercing way of clarifying what really matters, along with what does not.

Studies show that people who possess strong internal self-awareness make better decisions, have better personal relationships, are more creative, and have more fulfilling careers. It is also associated with improved mental health and general well-being. Internal self-awareness allows us to abandon balance but without being swept away by rote inertia or forfeiting our agency to *choose* how we spend our time, energy, and attention.

When you put all of this together, an empowering idea begins to emerge: Excellence is not about trying to achieve some sort of illusory balance. Instead, it's about pursuing our interests wholeheartedly while maintaining the ability to see ourselves clearly and make adjustments if necessary.

What About Obsession?

A common narrative is that in order to excel at something, you have to be maniacally obsessed—you must eat, sleep, and breathe your activity to the exclusion of all else. It is simply the cost of greatness. People point to athletes like Michael Jordan or Tiger Woods to illustrate the idea. Yet the data tell a different story.

Greatness requires you to make sacrifices, try hard, overcome pain and discomfort, and spend an outsized proportion of your time and energy on your primary pursuit. There is no way around it. You may even need to be obsessed for specific periods of time. But chronic obsession—becoming completely fused with an activity; thinking about it all the time; going on autopilot and feeling like you have to do it even when you don't want to—is almost always a recipe for disaster.

Yes, there are some exceptions. But research shows that for the vast majority of people, including those at the very top of their fields, chronic obsession results in decreased performance, burnout, and unethical behavior. There are countless stories of people who had the potential for greatness and whose careers (and sometimes lives) were ruined because of their obsession. They suffered poor physical and emotional health and often cheated or engaged in fraud. I'm sure you can recall plenty of examples. When an activity becomes all we do and all we are, when the stakes constantly feel like do or die, we find ourselves in a precarious position. The name of the game is harnessing the craziness, obsession, and drive and putting some constraints and boundaries around it without losing the fire. It's a never-ending path and the work of a lifetime.

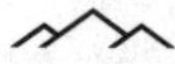

There are plenty of lifestyle and happiness gurus who espouse balance above all else. There are also plenty of motivational speakers who assert that you must always be obsessed to succeed. But the truth is more nuanced.

In this context, balance means you *never* go all in. Obsession means you *recklessly* go all in. Prioritization means you go all in but not on autopilot and not in a way that ruins your physical, emotional, or relational health.

When you zoom in on any one period of an accomplished person's life, they don't appear to be balanced. But when you zoom out and look across the entirety of their life, they seem to be quite balanced. They give themselves permission to go all in on the activities they care about, making challenging trade-offs along the way. At the same time, they remain in touch with the other important areas of their lives. In my years of researching and reporting on high performers, I've observed this theme in individuals pursuing their own versions of excellence in nearly every domain. They have different seasons for emphasizing different pursuits. It's a model that all of us can follow.

Keep the Main Things the Main Things

Everyone knows the person who wants to be an excellent athlete, and so that aspiring athlete takes fifteen different supplements, cold plunges every morning, and tracks their biometric data. But they don't train regularly, eat their fruits and vegetables, build community, or relax. They obsess over the 0.1 percent but not the 99.9. This pattern occurs across domains, from sports to the arts to knowledge work. People do all sorts of bizarre things to "optimize" their performance without consistently showing up and nailing the basics.

Two types of prioritization are vital to excellence. The first, which we covered in the preceding pages, is prioritizing within the broader context of your life. It's about protecting the time and energy to give your all to what you care about, without spiraling into reckless obsession and burnout. The second, which we'll turn to now, is about prioritizing what you focus on *within* your primary pursuit.

If you consume information about fitness, music, productivity, or generalized performance, then you know there is an endless stream of high-tech gadgets and fancy protocols on offer. It's even true for something as basic as breathing, which has become a billion-dollar industry thanks to various beliefs about the best ways to inhale and exhale. We've become obsessed with complexity. Sometimes this is warranted, but often it is not.

Complexity is a way to avoid facing the reality that what really matters for progress in most endeavors is simply showing up and doing the work. The more complex you make something, the easier it is to procrastinate. Complexity gives you excuses, ways out, and endless options for switching things up all the time. It gives you plenty to think and talk about, but also too much to do, at least with any real consistency. The result is that you all too easily bounce from fad to fad.

Simplicity is different.

You can't hide behind simplicity. You have to show up—day in, day out—and do the work. People love bright and shiny objects, but the truth is that the way you improve in anything is by eliminating distractions, prioritizing the fundamentals, and executing them with ruthless consistency.

Many record-breaking runners train on dirt tracks and eat simple diets. Many bestselling authors still write at the same intervals and at the same desks at which they started. Too often we think a great feat was accomplished *despite* the approach being basic and uncomplicated when in fact it was accomplished *because* the approach was basic and uncomplicated.

You can get amazingly strong with one kettlebell. You can write beautifully with a pen and a notebook. You can develop a spiritual practice with nothing but your breath. It really is as simple and as hard as that.

Dan John is a world-class strength coach. For the past forty years, he's worked with countless Olympians and professional athletes. About a decade ago, he started to notice an interesting trend. People would show up at the gym or track and proceed to spend more than an hour engaging in a wide variety of warm-ups, drills, and accessory exercises—the majority of which were fads that cycled in and out with regularity. But when it came time for the main movements—the stuff that matters, the stuff that hasn't changed in the last two hundred years—they were worn out. They struggled to focus and rushed to leave the gym so they could get to their so-called recovery work: ice baths, blood glucose monitoring, and on and on.

John observed that these same people then wondered why their improvement had stalled despite obsessing over all the latest and greatest protocols. The problem, he came to say, is that they weren't keeping the main things the main things.

It's not just sports. Whether it's cooking, photography, music, research, craftwork, gardening, painting, or something else, it's never been easier to get lost in the minutiae and details of your activity at the expense of the main things. But this is a mistake.

Every life and every pursuit has a foundation, and every life and every pursuit has accessories and auxiliary activities. The foundational work tends to stay the same. The accessories and auxiliary activities tend to be new and exciting. But don't confuse which of the two matters more.

Regardless of your activity, it is crucial to define the fundamentals—the main things. Also define the temptations, the bright and shiny objects, the trends and fads. Your job is to make sure that you are focusing on the former and not wasting your time, energy, or attention on the latter.

It's true that the best performers check all the boxes, but they also make sure that all the boxes are worth checking in the first place. You've got to ensure you aren't worrying too much about the last 0.1 percent and not enough about the first 99.9.

It's a good practice for our lives as a whole, too.

Chapter Summary

- Balance is an illusion.
- Prioritize what you care about.
- Self-awareness allows you to evaluate trade-offs and choose where to spend your time and energy.
- Going all in can be a wonderful source of fulfillment so long as you don't completely leave behind other important parts of your identity—minimum effective doses can help.
- Think about having different seasons of life for prioritizing different pursuits.
- Whenever possible, keep it simple and be wary of hiding behind complexity.

Don't major in the minors.

Keep the main things the main things.

8

Focus

Algorithmic mass distraction: the experience of being served up an interruption engineered to capture your attention for a short period of time, before being served up the next interruption engineered to capture your attention for a short period of time, ad infinitum.

Focus: concentrated and enduring effort pointed toward a specific aim.

There is an ongoing battle between algorithmic mass distraction and focus. Excellence requires winning it.

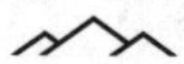

Another noteworthy term is *attention economy*. It describes the monetization of distractions that capture our attention. If someone can capture our attention, even if only for a few seconds, they can earn a profit from advertisers who want to sell a product or service. The incentives are astounding. As of 2023, economists estimated the global attention economy to be trillions of dollars, explaining why there is no shortage of novel and irresistible content filling our screens and airwaves at all times.

The rise of the attention economy, and the algorithmic mass distraction it spawns, has coincided with the rise of a culture that champions optimization, the idea that you can get more done and consume more information in fewer hours. In the majority of self-improvement and biohacking circles, the goal is not to be fulfilled, effective, or even happy, let alone excellent. The goal is to be *optimized*. It is a perfect storm, the result of which is overload and distraction cloaked as productivity and abundance. There is perhaps no better example than multitasking.

When we multitask, we do (and nowadays also consume) multiple things at once. Sometimes we multitask voluntarily. Other times, we multitask due to frequent interruptions that we didn't ask for. Often it's a combination of the two. Multitasking makes us feel productive because we believe that we are accomplishing twice as much. Taking a call at the same time we are in an online meeting at the same time we are writing an email; watching a television show at the same time we are reading an article at the same time we are scrolling social media at the same time we click over to buy a pair of shoes—this is achieving optimization, or so we tell ourselves. Like an advanced computer, we are running several processes at once.

But our brains are not computers. Effective multitasking is really nothing more than effective delusional thinking.

Even in individuals who claim to be great multitaskers, fMRI scans of the brain reveal it is impossible to do two things at once with a high level of quality. What is happening when we think we are multitasking is that our brain is either constantly switching between tasks or it is dividing and conquering, allotting only a portion of our cognitive capacity to any specific activity. According to countless studies, the consequence of multitasking is that the quality and, ironically, even the quantity of our work suffers.

The costs of task-switching may seem trivial, sometimes requiring only a tenth of a second to go from one tab to another on a computer. But over time they add up. When you switch from task to task, it's not just the physical act of shifting your gaze that consumes attention and energy but also the attention and energy required to transition your thinking and feeling.

My colleagues at the University of Michigan demonstrated that even seemingly innocuous multitasking, such as looking over to an on-screen advertisement while reading an article, can cannibalize as much as 40 percent of someone's productive time. A study conducted out of King's College in London found that persistent interruptions in our attention lead to a ten-point drop in IQ, or about twice the decrease one experiences after cannabis use and on par with what you'd expect after staying up all night.

Other research shows that scattering our attention also negatively impacts our emotional health. A study from Harvard University found that when people are fully present for what they are doing, they are much happier and more fulfilled than when they are thinking about something else. The more diffuse our attention, the more likely we are to feel angst, restlessness, and discontentment. "A wandering mind is an unhappy mind," conclude the researchers.

Multitasking detracts not only from how we feel and what

we do today but also our tomorrows. Over time, people who are chronic multitaskers become worse at filtering out irrelevant information and slower at identifying patterns, and they experience a decline in long-term memory. Not to mention, constantly switching and fragmenting our attention exhausts and depletes us, leaving us with less energy for what actually matters in life. If we fry our brain on the trivial, we cannot expect it to be fully present for the significant.

Unfortunately, for many people, fragmented attention has become the default way of existing. Research from the United Kingdom's telecom regulator shows that the average person checks their phone every twelve minutes, and this doesn't include instances when someone thinks about checking their phone but does not. Other research shows 71 percent of people *never* turn off their phones and 40 percent look at their phone within five minutes of waking up, and that's apart from the alarm clock app.

The consequence of all this checking is that we are habituating ourselves to distraction.

Gloria Mark, an attention researcher at the University of California, Irvine, and author of the book *Attention Span*, began studying people's ability to focus in 2003. Back then, the average duration that someone could maintain deep attention was about two and a half minutes. Today, she says, that number is forty-seven seconds, and it's only getting worse.

We've been conditioned to believe that if we aren't constantly being entertained or excited, if we aren't looking back or thinking ahead, then something must be wrong. But the opposite is true: When we scatter our attention, when we are constantly interrupted and distracted, we become alienated from whatever it is we are doing. Excellence, and the deep satisfaction it brings, requires the ability to sustain focus. And the ability to sustain focus is increasingly becoming a competitive advantage in today's world.

Eliminating Distractions

There are myriad ways to build focus. Chief among them is mindfulness meditation. It involves attending to a single anchor, most commonly the breath, for a specified period of time. Whenever you get distracted—be it by a thought, sensation, urge, or sound—you simply notice the interruption and then return to focusing on your breath. Do this regularly, even if for only ten minutes per day, and you begin to develop your attention muscle.

Many people benefit tremendously from mindfulness meditation, myself included. But I fear we miss the forest for the trees when we focus solely on what is happening inside our head and not everything that is happening around it. If you take even the most skilled meditator, give him a smartphone or laptop, and then ping him with notifications that have come to represent his individual worth, significance, and safety—comments, likes, shares, emails, and news alerts—then he, too, would probably struggle to sustain attention. There is a reason monasteries tend not to have active 5G connections.

In my prior work, I've used a handful of metaphors for the powerful devices that we carry on our person. The two most popular are "an open bag of candy" and "an existential slot machine." The former represents the infinite well of ultra-processed content, the consumption of which makes us feel good in the moment but leaves us empty later on. The latter represents the chance to swipe down (instead of a pull lever) for the opportunity to be rewarded with a notification symbolizing our significance in the world. Both are scientifically constructed to exploit primitive networks in our brain that are associated with addiction.

We know from decades of research that placing someone with a binge eating disorder in front of an all-you-can-eat buffet or taking someone with a gambling problem to a casino is a recipe for disaster. And yet this isn't far off from what we do to ourselves every time we attempt to focus in the presence of digital devices. Even if we don't constantly check them, the amount of cognitive and emotional energy required for restraint is draining.

In a series of studies published in *The Journal of Social Psychology,* researchers asked participants to complete difficult tasks when their smartphones were visible. Sure enough, performance was worse than in a control setting in which participants' smartphones were not visible. However, the interesting twist occurred in a third condition, where there was a smartphone present but it did not belong to any of the study participants. Rather, it belonged to one of the researchers. Still, the participants' performance declined. Even if it is not our own, the mere sight of a smartphone has come to signify everything else that could be happening in the world. Just being in its presence is a powerful source of distraction.

Athletes neither train nor compete with their smartphones in their pockets, at least not when they are at their best. The same can be said for writers who place their digital devices in another room when they sit down to write. Great artists, craftspeople, and researchers do the same.

The most effective strategy for deep focus is identifying and removing distractions. Full stop. It may not be easy, but it sure is simple.

Distractions are an inevitable part of our lives, so while it is unrealistic to eliminate them altogether, we can identify specific key activities, set aside time and space for them, recognize sources of distraction, and physically remove them. I've come to call these *deep-focus blocks*, and I have found they are crucial for sustaining excellence as well as promoting attentional health.*

A common pushback to deep-focus blocks comes from people who use digital devices for their work and thus cannot physically remove them. Examples include nonfiction writers like myself, software engineers, designers, architects, attorneys, and so on. They say, *Closing my browser, exiting my email client, and turning off the internet all sound great, but it is so easy to turn back on.*

Yet there are still steps you can take. For starters, you can commit to disabling distracting functions and apps during deep-focus blocks for two weeks, and see how you do. You may be surprised at the effectiveness of even seemingly slight barriers to entering algorithmic mass distraction. If that doesn't work, you could consider acquiring a second device that you use only for deep-focus blocks. Strip this device of potential distractions and make it as bare-bones as possible—never connect it to your Wi-Fi, for example.

You may not find a perfect solution, but you can continually ask yourself, *What is coming between me and intimacy with what I'm doing, and how can I best remove those barriers and distractions?*

---

* Here I want to credit Cal Newport, who coined the phrase *deep work* in his seminal 2016 book by the same title. The idea of deep-focus blocks grew out Newport's work.

Executing Deep-Focus Blocks

Most people require at least a few minutes to settle into focus. For example, when I sit down to write, I always face some initial friction, restlessness, and irritation. I've learned to accept this as part of the process, stick with what I'm trying to do, and let the resistance pass. This is true for nearly all forms of deep-focus work, regardless of the pursuit. You've got to get past the resistance for the good stuff to happen. Research from across diverse fields demonstrates that the ideal window for attentive engagement ranges from about fifty minutes to two hours. Anything less and you don't give yourself enough time to get into a rhythm. Anything more and fatigue sets in.

It's crucial to be patient, especially at first.

Even if you remove external distractions, the mind still bounces from one thought to the next. It can take a few weeks to rebuild your attentional capacity. When someone who hasn't read a hardcopy book for quite some time attempts to sit down and read, even if they've read thousands of books in the past, it is inevitable that at first they'll struggle. But after a few weeks of regular practice, they'll find it much easier to settle in. I know this firsthand. There are (too many) times when I succumb to algorithmic mass distraction and enter a reading rut. Whereas just a month before, I could read deeply for two hours at a time, now I can't make it through a single page without becoming distracted. But I know that if I remove digital devices from the room I am reading in and commit to the process, after a week or two of regular practice I'll find it easier to become absorbed in a good book again.

My experience with reading represents a predictable pattern for reclaiming focus in anything. Focus is like a muscle: Use it or lose it; even if you were once strong, regaining your fitness takes time and effort. But with patience and repetition, it comes back.

You may also benefit from keeping a physical notebook nearby during deep-focus blocks. If a thought surfaces that is unrelated to what you are doing and yet still feels memorable, you can jot it down and move on. The advantage of a *physical* notebook ought to be clear: It is not a digital device, and thus it is not a gateway to endless distraction.

It's important not to judge yourself or panic when you experience internal interruptions. Decades of psychological research show that the more you resist thoughts, feelings, or urges, the more persistent they become. It's best to simply notice them, write them down if necessary, and then return to the object of your focus. Do this over and over again, and your focus improves.

At some point, even if you've done everything above, your attention is bound to wane. If it's been over fifty minutes and you reach an impasse, that's a sign you may be ready for a break.

Research shows that breaks ranging from five to thirty minutes are best for restoring attention. The longer the preceding period of focus, the longer the break. One of the most effective breaks is a short walk. In an ideal scenario, you'd do this outdoors, but you could also walk around your office or home.

The idea for this book first occurred to me on a short walk, one I embarked upon because I was stuck on an unrelated project. When the idea came to me, I was so excited I couldn't wait. I remember calling my literary agent from the Hominy Creek Greenway and telling her, "I think I can build upon some of Robert Pirsig's ideas and take a big swing at a book on excellence."

She said, "Tell me more," and here we are.

What makes this story remarkable is how common it is. Many people I know have had truly life-changing ideas during walks. The philosopher Friedrich Nietzsche once said that "all truly great thoughts are conceived by walking." Often we think we are taking a break from productivity only to have the break turn into the most productive part of our day, week, or year.

In a study aptly titled "Give Your Ideas Some Legs," researchers from Stanford University found that individuals who took a six-to-fifteen-minute walking break increased creative thinking by 40 to 60 percent as compared to those who remained seated at a desk. At first the researchers speculated that increased blood flow to the brain was responsible for the benefits of walking. But it appears the benefits might also emerge from the interplay between walking and attention: Mainly, walking requires just enough coordination to occupy the parts of our brains that control effortful thinking, allowing us to zone out and mind-wander more easily. Even though we are using the effortful thinking networks, we are doing so in a down-regulated way, in essence giving them a chance to rest.

If you are unable to walk, then listening to music, looking through a window at nature, taking a shower, doing the dishes, or preparing a meal are also good options. The goal of these breaks is to switch from whatever it is you were doing to something that gives your deliberate thinking and attentional capacities a chance to rest.

How many blocks of deep-focus engagement should we aim for?

As many as possible.

For some of us, this may mean three blocks per day. For others, it may mean three blocks per week. What is realistic will depend on your primary pursuit and the other demands of your life. Either way, it is remarkable how much better you feel and do with a few deep-focus blocks.

Schedule deep-focus blocks on your calendar, lest they all too easily get lost in the commotion of life. They are a meeting with yourself and your most important work, which ought to make them the most important meeting there is. Even better, make deep-focus blocks a part of your regular routine. For instance, you could set aside Monday, Wednesday, and Friday from 8:30 to 10:00 a.m. What matters is proactively creating and protecting the time and space to step outside of algorithmic mass distraction and enter into circumstances that promote intimacy with meaningful projects and pursuits.

It's no surprise that we feel content and fulfilled after a day that includes deep-focus work. Even when the effort is hard, it is still good. Our world may be replete with distraction and noise, but excellence—and the deep satisfaction that accompanies it—demands a commitment to strengthening our signal.

Chapter Summary

- It is harder than ever to focus, in large part due to attention economy incentives.

- Multitasking is an illusion. We are at our best when devoting our full attention to the task at hand.
- Focus is like a muscle. It takes time to build. If you don't use it, you lose it.
- Don't expect deep focus to happen on its own. Schedule blocks of time for it, and remove your digital devices and other sources of distraction.
- The ideal window for attentive engagement ranges from about fifty minutes to two hours, followed by five-to-thirty-minute breaks.

The quality of your attention
shapes the quality of your life.

9

Discipline

Layne Norton is a world-champion powerlifter. He holds the record deadlift for his division, an astonishing 723 pounds. It took Norton twenty years of serious training to achieve that mark. When I asked him about his stick-to-itiveness, he told me it comes down to "disconnecting how you feel from what you need to do. There are times when you are going to feel very motivated, and that's great. But discipline is: You are going to do the things you need to do, regardless of how you feel about them . . .

"As humanity has advanced, we have been able to devote more space for our feelings and indulge them more. That isn't necessarily a bad thing. It has produced more compassion and empathy, and those are good things," continues Norton, who also holds a PhD in nutritional sciences. "But in some ways, we have given too much space to our feelings. Many people end up completely governed by feelings, and for them, life can be pretty hard."

Your emotions can change on a dime and often not for any good reason. It takes self-reflection to sit with emotions and ask whether or not they are reflecting truth or something else—*before* acting on them. But many people believe that just because they feel something, it must be the same thing as truth.

One of the most common questions Norton receives is "How do you stay motivated?" "The answer is I don't," he says. "But I also don't worry about being motivated all the time. I often do not feel like training. Sometimes I don't feel like working. Sometimes I don't feel like brushing my teeth. But I still do all those things consistently. Why? Because I know that if I don't do them, I will not achieve my goals. So I do them regardless of my feelings. I'm not saying to not have feelings. I'm not saying to not pay attention to them. I'm simply saying that sometimes you need to disconnect your immediate feelings from the things you know you need to do in order to reach your goals and build a great life."

A common misconception is that the best performers are always motivated and inspired. That couldn't be further from the truth. Olympians, award-winning artists, and successful entrepreneurs do not wake up every morning raring to go. They struggle just as much as the next person not to hit the snooze button.

I know this struggle, too.

If I had to be motivated every time I sat down to write, there would be hardly any writing. If I had to feel ready to crush it every time I trained, I would have done eighteen workouts last year, not 260.

We think we need to feel good to get going, but often the opposite is true: We need to get going to give ourselves a chance at feeling good. This is where discipline comes into play—a commitment to consistently showing up requires it.

Activation Energy

In the late 1970s, the clinical psychologist Peter Lewinsohn noticed that just because he uncovered an uncanny insight from a patient's childhood or identified irrational thinking patterns, it did not necessarily change their mood. What made a difference was supporting his patients in taking productive actions. If he could help them get started on small tasks even when they didn't feel like it, their moods had the best chance of improving. The tasks could be simple: doing the laundry, going for a short walk, calling a friend. It was a departure from the prevailing therapy wisdom at the time, which centered around driving change by unlocking insights and shifting negative thought patterns.

In 1974 Lewinsohn published his findings in a groundbreaking paper, "A Behavioral Approach to Depression." Since then, many studies have demonstrated that controlling our thoughts and feelings is hard, if not impossible. The more we try to suppress or change a certain thought—for instance, *I really don't want to get started today*—the stronger that thought becomes. The same holds true for emotions. When we try to force ourselves to feel differently, we're liable to get even more stuck in our current state.

What we can control, however, is our behavior—that is, our actions. And if we take action, we can change our mood and mind. Action can *create* motivation.

Nowadays, Lewinsohn's method is called *behavioral activation*. It is used widely in clinical psychology where it is part of a broader tool kit for treating depression and anxiety. But its core premises—that motivation and inspiration are merely sensations that come and go, just like any other feelings; and that we can *act* our way to new emotional states—are relevant for us all. In moments when motivation is lacking, our self-talk can be as simple as *I'm feeling a bit flat right now, but that's okay. Let's try to nudge into action and see what happens as a result.*

There are things we can do to support motivation and inspiration, such as reading good books, spending time in nature, and surrounding ourselves with energizing people. But ultimately motivation and inspiration operate on their own schedules, so there is no point of becoming beholden to these states. When we are down, unmotivated, or apathetic, we can feel those feelings but not dwell on them or take them as destiny. Instead, we can shift the focus to getting started with what we have planned in front of us, taking our feelings, whatever they may be, along for the ride. If we get going and still feel off, we can always stop. The key is that we get started and give ourselves a chance. It's what discipline is all about.

"Don't wait for the muse," exclaims the bestselling author Stephen King in his memoir, *On Writing*. "Your job is to make sure the muse knows where you're going to be every day from nine 'til noon, or seven 'til three. If he does know, I assure you that sooner or later he'll start showing up."

King is not a master of motivation or inspiration. He is a master of discipline.

The challenge is mustering enough energy to get going on the things that matter to us, even when we don't feel like it. It can help to think of the initial oomph as *activation energy*. The amount of activation energy we require can change from day to day, week to week, and even year to year. Sometimes we'll feel motivated and inspired, requiring little activation energy. Other times we'll be faced with resistance and friction, and we'll need to rely heavily on activation energy, which is precisely why discipline is so important.

Discipline bridges the gap between motivation and action, making the former less necessary for the latter. When you have discipline, you don't need to feel a certain way to show up and get started. You just do.

Discipline is not a chest-thumping, performative act of toughness. It is being the kind of person who shows up for what matters and does what you need to do. The irony is that when you do hard things that you don't feel like doing in the short run, you usually end up feeling better in the long run.

The Freedom to Be Your Best

The powerlifter Layne Norton isn't the only world-champion to understand the power of showing up and discipline. The marathoner Eliud Kipchoge and the tennis player Venus Williams are two of the greatest athletes to ever live. Both compete in sports that require relentless consistency. Yet neither Kipchoge nor Williams are world-class at motivation. Each frequently undergoes periods of training that feel monotonous and tedious.

When asked about his success, Kipchoge says, "Only the disciplined ones in life are free. When you are undisciplined, you are a slave to your moods and passions."

Williams explains, "I honestly believe discipline is freedom."

Kipchoge and Williams are alluding to what philosophers call *positive freedom*, or the freedom to express your potential. But in order for that to happen, you need to sacrifice what philosophers call *negative freedom*, or freedom from constraints.

When you cultivate discipline, you are placing a constraint on whether or not to get started. You don't have to think about getting started. You don't have to feel a certain way to get started. You just get started.

There are other constraints, too. Many who are committed to excellence live relatively structured lives. They practice their respective crafts, take care of their minds and bodies, and spend time with the people they care about. Someone like Kipchoge could take advantage of his celebrity status with endless partying and gala appearances, yet he chooses not to. "I always tell people that this is a really simple deal: Work hard. If you work hard, follow what's required and set your priorities right, then you can really perform without taking shortcuts. If you're taking shortcuts, you can't be free," he says.

The number of constraints required for excellence depends on your unique temperament and what you are trying to accomplish. Too many constraints and the result is rigidity. Too few constraints and the result is instability.

You could begin by identifying the key junctures in your respective craft. For an athlete, this might mean starting your workout every morning at 6:30 a.m. For a physician, it could mean showing up at the clinic with a present mind at 8:00 a.m. For a writer, it might look like facing the blank page from 7:00 a.m. to 11:00 a.m.

Next, you want to explore the levers that occur upstream of those crucial moments. What detracts from your activation energy, and how can you do less of it? What bolsters your activation energy, and how can you do more of it? What constraints—perhaps on how you eat, drink, sleep, consume information, and otherwise spend time—might support your pursuit of excellence?

This sort of discipline is not for everyone. But if you commit to it, the rewards are substantial. "When you have discipline you are free to live your dream, free to do the things you love, and then you come to love the discipline," says Williams.

Never Prejudge Performance

There is a scene in the Beatles documentary *Get Back* that begins with the band stumbling into a recording session with hardly any energy. Paul McCartney is visibly frustrated, noting that "Lennon's late again." George Harrison is yawning and struggling to keep his eyes open. Ringo Starr appears exhausted and zoned out. Nobody wants to be in the studio. Nevertheless, McCartney lazily begins strumming an A chord on his bass, gibbering ad-lib lyrics as he goes. Eventually he lands on the words "get back." Those two words have an immediate effect on Harrison and Starr, who return to life. Lennon finally enters the studio, grabs his guitar, and joins in. Herein lies the genesis of one of the most iconic songs ever written.

Imagine if the Beatles had decided not to get started that day.

Or if McCartney had referenced a wearable that, based on a black-box algorithm, told him in order to "optimize" his performance, he would need to stay in bed and rest.

Or if the band had told themselves, *Today just isn't the day,* and created a self-fulfilling prophecy around that thought.

Another type of discipline that is crucial to excellence is the discipline not to prejudge performance. There are days when I feel great and highly motivated to write, yet when I sit at my computer, nothing comes of it. There are days when I feel dreary and apathetic, but once I get going, I produce some of my strongest work. I've learned that it's hard to predict which direction a writing session will go, so the best bet is always to get started and give myself a chance.

Zach is one of the better athletes I know. Not only is he fit in his own right, but he's also a highly sought-after coach with a degree in exercise science and more than fifteen years of experience. He understands the body as well as anyone. In his own training, Zach had been focused on cycling. He planned to test his progress on Elk Mountain, an all-out climb that involves nearly five miles of unforgiving ascent for 1,300 feet of elevation gain. But on the morning of his test, he felt "flat."

He wasn't ill or injured, and yet he wasn't sharp either. From his years of studying performance and working with athletes, Zach knew that how you feel *before* a hard effort does not necessarily correlate with how you feel *during* a hard effort. So he took his not-so-great feelings along for the ride (in this case, literally) and ended up smashing a personal record, laying down one of the best performances of his life. At no point did he predetermine what would happen. Instead of spiraling into negativity when he felt off upon waking up, he mustered the discipline to show up, give himself a chance, and get on with the show.

For all that we know about how our minds and bodies work, much is still mysterious. Human performance is complex. We can always shut things down if need be, but if we have the discipline to keep an open mind, get started, and give ourselves a chance, we never know when we'll be pleasantly surprised.

Fierce Self-Discipline Requires Fierce Self-Kindness

If you've ever browsed at the bookstore, you've probably noticed a section of self-discipline books that tend to be written by Navy SEALs, marines, and football coaches. These books extol the virtues of hard work, personal responsibility, and showing up under all circumstances. They remind us that life is hard and that nobody will do our bidding for us.

You may also have noticed a section of self-kindness books. These books tend to be written by meditation teachers and therapists. They focus on boundless love and softening our hearts. They also remind us that life is hard, but suggest it's all the more reason to go easy on ourselves.

While these two philosophies—self-discipline and self-kindness—are often pitted against each other, the truth is they go hand in hand. If we are always beating ourselves up, we will not last. But if we never push ourselves, we won't ever attain our full potential. Even the most excellent people endure doubts, intrusive thoughts, and periods of pain and apathy. The work is accepting these thoughts and feelings and showing up as best we can nonetheless. Sometimes the kindest thing is also the hardest—and what seems hard in the moment is the key to creating a better future.

Combining self-discipline and self-kindness is a profound mindset shift that is essential to sustaining excellence. It's the difference between self-talk that sounds like . . .

You're never motivated. You always feel like crap. Why is everything so hard for you? You better get going or else you are going to be a failure . . .

. . . and self-talk that sounds like . . .

What you are trying to do is hard, but you are capable of doing hard things. This matters to you, and nobody is going to do it for you. Let's muster some gentle yet firm persistence, get started, and see what happens.

Showing up is hard.

Setting constraints is hard.

Delaying gratification is hard.

All of which is to say that maintaining discipline is hard.

It is important to remember that you can do hard things. But it is also important to remember that doing hard things becomes just a bit easier—and far more enduring—if you can learn to have your own back and be your own friend. Fierce self-discipline benefits from fierce self-kindness. These qualities are not opposites, they are complements. If you want to be an extraordinary badass, you've got to be kind to yourself, too.

Chapter Summary

- Discipline bridges the gap between motivation and action, making the former less necessary for the latter.
- You don't need to feel good to get going; you need to get going to give yourself a chance to feel good.
- The greats are not masters of motivation or inspiration; they are masters of showing up and getting started. They are masters of discipline.
- Discipline demands sacrificing negative freedoms *from* constraints for the positive freedom *to* be your best.
- Too many constraints and the result is rigidity, too few constraints and the result is instability.
- Never prejudge performance.
- Fierce self-discipline requires fierce self-kindness.

Discipline is not a chest-thumping performative act of toughness.

It is being the kind of person who shows up and does what you need to do.

10

Renewal

During a 1958 interview with George Plimpton for *The Paris Review*, the writer Ernest Hemingway explained that he'd often begin working as early as six in the morning but would finish no later than noon. "You write until you come to a place where you still have your juice and know what will happen next and you stop and try to live through until the next day when you hit it again," he said. "It is the wait until the next day that is hard to get through."

Just as it takes discipline to show up and get started, it takes discipline to rest. This is especially true in a society that glorifies grinding, short-term gains, and pushing to extremes. When Plimpton asked Hemingway if when he stepped away from the typewriter he was able to remove his mind from whatever project he was working on, Hemingway replied, "Of course. But it takes discipline to do it and this discipline is acquired. It has to be."

I can empathize. Perhaps you can, too.

It is hard to step away from the projects that we care about and identify with. During the process of writing this book, on numerous occasions it was physically painful for me to set down the manuscript and head out for a walk or drive to the gym. Every cell in my body urged me to keep working, to continue focusing on whatever problem I was trying to solve. Yet, like so many other writers before me, I've learned how critical it is to step away, even if it feels like I'm forcing myself.

Consistent progress requires periods of rest and renewal.

It's when our bodies rebuild and repair; when our minds become more creative, replenish willpower, and regain emotional control. We'd be wise to think of rest and renewal not as something we do at the expense of our primary activity but rather as an integral *part* of our commitment to sustaining excellence.

"A good idea doesn't come when you're doing a million things," says Lin-Manuel Miranda, the creator of the blockbuster Broadway musicals *In the Heights* and *Hamilton*. "The good idea comes in the moment of rest. It comes in the shower. It comes when you're doodling or playing trains with your son. It's when your mind is on the other side of things . . .

"When I picked up Ron Chernow's biography [of Alexander Hamilton], I was at a resort in Mexico on my first vacation from *In the Heights,* which I had been working seven years to bring to Broadway. The moment my brain got a moment's rest, *Hamilton* walked into it. It's no accident that the best idea I've ever had in my life—perhaps maybe the best one I'll ever have in my life—came to me on vacation," he adds.

There is no free pass to excellence; we need to show up regularly, challenge ourselves, and push our limits in order to improve. But doing so is only effective when our efforts are accompanied by rest and renewal. Too much stress, not enough rest, and the result is injury, illness, exhaustion, and burnout. *Stress plus rest equals growth.*

Our muscles get bigger not when we are training but rather during periods of recovery between workouts. It is only then that the body adapts to the strain that was placed on it, growing stronger as a result. The same is true for our brains. It's why so many creative thoughts, breakthrough ideas, and aha moments occur not when we are actively working on an important task but during breaks—while walking, showering, driving, cooking, cleaning, or just upon waking. Researchers frequently describe creativity as a cycle of immersion, incubation, and insight. First, you throw yourself into the work. Then when you find yourself becoming exhausted and getting stuck, you step away. This gives you the best chance at unlocking a key discovery, and it explains why so many of our greatest ideas occur in the shower, while on a walk, while doing the dishes, upon waking from sleep, or during a long drive.

There is a science—and an art—to rest and renewal. Not all forms are created equal. Many of the activities we think are restful may actually leave us exhausted. Before we get into the best strategies for rest and renewal, we must first agree on a common definition of what these terms even mean.

Most researchers describe rest and renewal as physiological states during which your innate stress response, or sympathetic nervous system, subsides in favor of a more relaxed condition. Your heart rate and blood pressure drop, and your shoulders usually follow. Psychologically, rest and renewal are considered a shift from deliberate and effortful thinking—for example, straining to solve a problem or trying to figure out the best way to communicate a complex topic—to more tranquil states, often characterized by mind-wandering or zoning out.

Stress and rest are subjective. A seasoned marathoner might consider an eight-minute mile to be restful whereas that same pace may be stressful for a novice. A bibliophile might find reading one of the iconic Russian novels a form of renewal, whereas a less avid reader may consider it hard work. There are, however, three features that nearly all forms of genuine rest and renewal share:

1. **You are not exerting self-control.**

   As Hemingway said, stepping away from what you are working on might require discipline up front. But once you are in the midst of resting, it should, by definition, feel easy. Trying hard to rest defeats the purpose. For example, it is not restful to force yourself to breathe deeply in silence when you'd rather be listening to music or taking a walk. Once you settle into a restful state, you shouldn't have trouble staying there.
2. **You are not consciously thinking about work or triggering topics.**

   Though your subconscious mind may be connecting dots and problem-solving in the background, your conscious mind is not on your work; it is floating freely. Likewise, a restful activity should not trigger anxiety, such as watching cable news, trying to get through a backlog of emails, or doomscrolling on social media. Though these activities may not tax your brain and body in the same way as deep-focus work, they prevent you from fully recovering.

   Studies show that when athletes are on social media between sets in the gym, their performance suffers. I've had numerous conversations with coaches from all levels—high school, collegiate, Olympic, and professional sports teams—and they've all told me that their athletes' execution deteriorates when they check social media at halftime or in between heats. As a result, some teams are beginning to mandate that players leave their phones in a lockbox prior to entering the training facility.

My athleticism is not on par with the pros, but I've noticed a similar trend in my workouts. When I check social media or email between exercises, not only does my performance suffer but my enjoyment does, too. I perform (and feel) best when I leave my phone in the car and either zone out or chat with other gymgoers during rest intervals.

It's no different with writing. I am better off taking a short walk, doing the dishes, or folding laundry between blocks of work than I am staring into a screen. The former renews my attention and energy; the latter depletes it.

3. **You are not turning rest into work.**

A common trap is becoming consumed with optimizing your "recovery" so that every meal, shower, walk, and even sleep turns into something to measure and excel at. Too much focus on rest and renewal has the ironic effect of creating exhaustion. If you are stressing about recovery metrics, then it is wise to consider ditching them. Once rest begins to feel like work, it is no longer rest.

Overemphasis on tracking can be particularly harmful when it comes to sleep. A study in the *Journal of Clinical Sleep Medicine* found that a reliance on sleep trackers led to an uptick in stress, which in turn made it harder for people to sleep. "We realized we had a number of patients coming in with a phenomenon that didn't necessarily meet the classical description of insomnia, but that was still keeping them up at night. They seemed to have symptoms related to concerns about what their sleep-tracker devices were telling them, and whether they were getting good quality sleep or not," says Sabra Abbott, an assistant professor of neurology at Northwestern University and a coauthor of the study. "In other words, people were stressed out—and in some cases, their sleep was suffering further—because they weren't measuring up to their tracker's definition of 'good' sleep."

Bottom line: If a recovery tracker is helping you to rest more, that's great. But remember that these tools can produce ironic effects. What works, works—until it becomes the very thing that gets in the way.

Plenty of activities meet the above criteria for rest and renewal, but a few in particular are backed by a great deal of evidence and almost universally beneficial. We'll turn to those now.

Walking (or Other Light Physical Activity)

When Hemingway forced himself to step away from the page, he'd go on a walk and follow it with a half-mile swim—both were anchors of his daily routine. He's not alone. Einstein, Asimov, Beethoven, Freud, Faulkner, Kafka, Hobbes, Descartes, Thoreau, Woolf, Tolstoy, Hawthorne, Tchaikovsky, Darwin, Blake, Wordsworth, and Dickens all cite physical activity as being integral to their creative routines.

As we previously discussed, activities such as walking, swimming, yoga, easy running, and hiking require just enough coordination to occupy the parts of our brain responsible for effortful thinking, allowing us to more easily zone out and mind-wander, both of which are associated with creativity and insight. Research shows as little as six minutes can produce benefits.

While physical activity is a no-brainer for knowledge workers, what if you are trying to recover from more physically demanding work? Perhaps you are a carpenter, nurse, or professional athlete. Even then, movement is helpful, so long as you keep it relaxed. Numerous studies demonstrate benefits to what sport scientists call *active recovery*, a fancy way of describing walking, yoga, easy swimming, or light cycling. These activities not only help put the mind at ease (which in and of itself has a positive effect on the body) but they also increase blood flow and mobility, which aids in muscle and bone repair.

Hanging Out

In the late twentieth century, scientists discovered the ratio of the hormones testosterone to cortisol acts as a gauge of stress and recovery. Testosterone is associated with growth and rejuvenation, whereas cortisol is associated with stress and breakdown—so the higher the ratio, the better. Studies have found that following stressful periods—such as a competitive sporting event, public speech, elongated session of deep-focus work, or intense session in the artist's studio—individuals who relax with friends experience a much quicker rebound in their testosterone-to-cortisol ratio. Other research shows that social connection helps to shift our nervous system into a restful state and promotes the release of biochemicals that have anti-inflammatory properties, such as oxytocin and vasopressin.

The modern science is interesting, but it shouldn't be surprising. For the vast majority of our species' history, we worked hard during the day and then unwound around the fire in the company of our tribe at night. We relax and rejuvenate when hanging out with friends, colleagues, and teammates because it is what we evolved to do.

A profound irony of the recovery industrial complex—cold plunges, compression boots, hyperbaric chambers, red-light therapy, and so on—is that many of these activities are undertaken alone, cannibalizing time that could otherwise be spent in community. More often than not, it's worth prioritizing relationships and connection, the benefits of which are almost always greater than what is on offer from new technologies and bright and shiny objects. It's not to say that these tools and technologies never have a place, especially when carried out with others. But like we discussed before, excellence requires keeping the main things the main things.

Time in Nature

From da Vinci to Darwin, many of history's greats reported being inspired by nature. In an attempt to bring empirical rigor to the topic, the University of Michigan psychologist Marc Berman conducted a study in which participants embarked upon a series of challenging cognitive tasks, such as solving elaborate verbal and numerical puzzles. After finishing, the participants were split into two groups: One took a break in a secluded park, while the other took a break in a busy urban setting. On a subsequent series of challenging tasks, the participants who took their break in a natural setting outperformed those who took their break in an urban setting.

In a second experiment, Berman put participants through the same process just described. Only this time, rather than go outside, they were instructed to view pictures of either natural or urban settings for a mere six minutes. The result was no different. The participants who viewed pictures of nature recovered faster and performed better on their next bout of challenges.

The most likely explanation for these benefits is the biophilia hypothesis, made popular by the Harvard scientist Edward O. Wilson.

The biophilia hypothesis states that since we as a species grew up in nature, we are biologically programmed to be drawn to it. A longing for nature is in our blood: We're hardwired to feel at home and at ease not in the city, suburbs, office, or gymnasium but in nature. As such, when we spend time in nature, we relax our guard and give our bodies and brains the opportunity to rest.

More recent research, largely out of Japan, offers additional support for the biophilia hypothesis. There, scientists found that when we immerse ourselves in natural settings, we experience significant positive changes in various biomarkers related to stress. Spending time in nature reduces cortisol levels, diminishes fight-or-flight nerve activity, decreases blood pressure, lowers heart rate, and restores our attentional capacity. If you work or live in a bustling area, even just gazing out the window at greenspace or walking a few laps around an urban park can go a long way to help.

Sleep

Much has been written on sleep, and there's no point in rehashing all of it here. Yet I'd be remiss not to mention a few highlights, along with a compelling and somewhat paradoxical idea.

First and foremost: No form of rest is as powerful as sleep. During sleep, our bodies repair and grow; and our minds retain, consolidate, and connect all the information that we were exposed to throughout the day. Naps of between ten and thirty minutes may provide a boost in energy and creativity, but nothing can replace sleeping seven to nine hours every night.*

However, there are times when getting appropriate sleep is impossible. An example is when you are caring for a baby. Fortu-

---

* See appendix 3 for a list of evidence-based practices that support sleep.

nately you can, in fact, go for extended periods with poor sleep and still be okay in the long term. (If this were not the case, there'd be data showing that those who have kids live shorter and less satisfying lives than those who don't have kids. But no such data exists.) I wrote *The Practice of Groundedness* when my first child was between the ages of zero and two, and frequently waking me up during the night. Did I feel at my best? No. But I continued to show up and write every day. I produced work I am proud of, and I survived just fine.

In the summer of 2025, the golfer J.J. Spaun completed one of the best career turnarounds in modern sport by winning the US Open. In the span of eight years, he went from being ranked 584 and missing the cut in many big tournaments to capturing a major championship. But the night before he won, Spaun was awakened at 3 a.m. by his two-year-old daughter, Violet, who had fallen ill. She couldn't stop vomiting. Spaun ran to CVS to get medication while his wife tended to his daughter. He described the sleepless night as "chaos."

The next day, he outplayed the field and won the championship.

If Spaun had slept so poorly for the entire prior year, surely it would have adversely affected his performance. But a single night is just that, and we are less fragile than we think.

It can be helpful to think about sleep using three essential rules:

1. Do everything you can to prioritize sleeping seven to nine hours.
2. Do not freak out if for some reason you can't sleep seven to nine hours.
3. Do not let rule 2 become an ongoing excuse for rule 1.

Extended Breaks

The concept of a weekend traces itself back to the early 1900s. At first, the weekend was largely a social norm to accommodate both the Jewish (Saturday) and Christian (Sunday) Sabbaths, both religious days of rest. In 1926, Henry Ford introduced limits on work for his employees, in large part because he realized those who worked forty-eight hours per week were less productive and more prone to burnout than those who worked forty hours per week. Sixteen years later, the United States government passed the Fair Labor Standards Act requiring overtime pay for anyone working more than forty hours per week, thus formalizing the modern weekend.

Fast-forward to today, however, and few of us take a proper weekend or observe Sabbath, either religiously or symbolically. We continue to press forward with our big projects from the workweek or we add additional stressors in other dimensions of our lives. But when we neglect to rest on the weekends, the quality of our ensuing week suffers, which pressures us to work on the following weekend yet again. A vicious cycle is born.

A simple way out is to create the equivalent of a sabbath that works for you and do everything you can to observe it. Rest days are not only for elite athletes; they are for all of us. One option for knowledge workers is a digital sabbath, or turning off all your digital devices for a set period of time. For example, you could shut everything down from Saturday morning to Sunday morning. I've done this in my own life, and I've come to enjoy the ensuing downregulation of my nervous system and forced reprieve from work and current events. It also has the effect of elongating my days and heightening my presence for what is happening directly in front of me. The upshot is that I return to my creative work energized and bursting with ideas.

Whether you unplug for an entire weekend, an entire day, or

even just half a day, time off comes with immense benefits. Studies show that both the quantity and quality of our work improve following a rest day. Extended breaks also allow us to reconnect with ourselves and those who are important to us.

Finally, there are vacations. Research from the PEW Center found that over 45 percent of Americans don't use all of their paid time off. But it is worth remembering that Lin-Manuel Miranda first had the idea for *Hamilton* on vacation. It's why professors take sabbaticals, why athletes take up to a month off after big events, and why so many species hibernate during the winter. Following prolonged periods of exertion, we benefit from prolonged periods of rest and renewal.

The decathlon is perhaps the most arduous event at the Olympics, though "event" is a misnomer since it involves ten different ones, including the 100-meter dash, hurdles, shotput, long jump, and 1500-meter run. As such, the winner is crowned "the world's greatest athlete." Damian Warner won gold at the 2021 Games, after more than a decade of competing at the highest levels in sports. Warner attributes a portion of his success to taking multiple months off following major competitions. "I view taking time off as something that is a requirement," he says. "The decathlon and everything that's involved with it, and with competing at a high level, is super stressful, so I think it's important to take a step back and live like a normal person."

If the greatest playwright of a generation and the gold medalist in the decathlon can rest, then you can, too. Though maybe that is the wrong way to think about it: Warner and Miranda didn't accomplish what they did in spite of the time they took off. They accomplished what they did, at least in part, because of it. Remember, stress plus *rest* equals growth. If you want to push hard, you've got to recover, too.

Chapter Summary

- We cannot grow or make progress without periods of rest and renewal.
- It takes discipline to rest, especially in a society that glorifies grinding, short-term gains, and pushing to extremes.
- There are three defining features of restful activities:
    - You are not exerting self-control by trying hard to rest.
    - You are not thinking about work or engaging in activities that cause you stress or anxiety.
    - You are not turning rest into work by tracking or obsessing over your recovery.
- Research-backed options for rest and renewal include walking or other light physical activity, hanging out with friends, spending time in nature, sleeping, and taking extended breaks—ranging from occasional days off to vacations.
- The more stress in your life, the more you should try to offset it with rest.
- Rest and renewal do not come at the expense of your primary pursuit; they are integral parts of your primary pursuit and your commitment to excellence.

It's hard to step away from work you care about and identify with,

but rest and renewal are crucial to progress.

11

Confidence

During nearly all my big writing assignments, there comes a point when I doubt where I'm heading. Early in my career these moments were terrifying. I was being paid to produce good work, often on a deadline. Sometimes a magazine would only give me one week to complete an essay, and I'd find myself lost at the end of day four. But over the years, I've discovered that if I stick to my process, eventually the path forward emerges, seemingly on its own.

We all face junctures when we are unsure of where the work is going or maybe even if it's going anywhere at all: the three-week mark in a challenging project when we find ourselves clueless as to how to move forward; the block of athletic training when progress stagnates, leading us to question our entire approach; the moment deep into studying for an exam when suddenly nothing makes sense. These instances of feeling lost are inevitable. They are inherent to growth, and nobody escapes them. When they occur, those of us who are relatively new on our paths may become stressed or dismayed. Whereas those of us who are more seasoned tend to stay cool, calm, and collected. It's a difference you see in athletes, coaches, leaders, craftspeople, doctors, attorneys, architects—in pretty much all fields at the expert level.

The late Zen Master Thich Nhat Hanh described faith as "the confidence a farmer has in his way of growing crops. It is not blind. It is not some belief in a set of ideas or dogmas."

The legendary music producer Rick Rubin calls it experimental faith: "When we sit down to work, remember the outcome is out of our control. If we are willing to take each step into the unknown with grit and determination, carrying with us all of our collected knowledge, we will ultimately get to where we're going. The destination may not be one we've chosen in advance. It will likely be more interesting. . . . This isn't a matter of blind belief in yourself. It's a matter of experimental faith. Over time, as you complete more projects, this [faith] grows."

The blank page is still harrowing for me, the feeling of uncertainty still uncomfortable. But facing it is not as hard as it used to be. I've learned that being unsure of where I am going is not something to stress over. It is merely a predictable part of the creative process, a necessary challenge of getting comfortable with being uncomfortable.

"It took me twenty years, but now I've got faith that after I finish a good song that I'm happy with, another one will come out of me. I used to freak out, especially if upon completing a song my creative energy was drained and I was feeling blank," the singer-songwriter John Moreland explains. "But I've done it enough times to understand it's all part of the process."

This developmental arc is true for all of us, regardless of what it is that we do.

The more reps we put in, the more faith we gain in our respective endeavors, and in ourselves. It is not blind or delusional faith.

It is faith based on a concrete body of evidence, which is a powerful way to think about confidence.

Confidence Requires Evidence

Arrogant people are loud. They need to convince everyone, including themselves, that they are up to the task—the reason being they aren't sure if they actually are. Underneath arrogance you'll almost always find insecurity.

Confident people are quiet. They know they've got what it takes. There is no need to waste anyone's time or energy, including their own, hemming and hawing. Underneath confidence you'll almost always find evidence.

No amount of outward bravado compensates for a lack of inner evidence. Even approaches like self-talk only work if you have reason to believe what you are saying. There is only one way to gain genuine confidence: by doing the work to earn it.

One of my friends is a renowned sports surgeon. He operates on the best athletes in high-stakes situations. "Ninety-five percent of the time in the OR [operating room] things go smoothly," he once told me. "But the other 5 percent, when the unexpected happens and you have to control excessive bleeding, change your plan when an implant fails, or fix an injury that is more severe than anticipated—I live for that. It allows me to bring all my years of training and experience to meet the moment."

His confidence is based on decades of deliberate practice and experience. "When we are in training with our mentors, it is great to see the cases that are effortless and learn the steps of how things should be done. But the ones the trainee appreciates the most are the cases that don't go well because then we see the masters at their best and in turn learn for ourselves how to deal with those complex situations in the future," he continued. "The most valuable cases that I saw in my training were the ones that went sideways and the best surgeons remained calm and knew how to get out of trouble."

The ability to remain calm amid challenges is a core element of what psychologists call *self-efficacy*: an evidence-based belief that you are capable of showing up, working through challenges, and excelling in uncertain or highly charged circumstances. Decades of research show that individuals who score high on measures of self-efficacy are better able to work through the inevitable moments when they feel lost or stuck, be it in the operating room, on the playing field, or in the boardroom. If you are insecure about your process and abilities, then you're liable to catastrophize when the path forward is unclear. But if you are secure about your process and abilities, if you have evidence to lean on, then not much can faze you. The best way to gain self-efficacy, the research shows, is through experience.

In a 1987 interview for her popular daytime show *Mavis on 4*, Mavis Nicholson asked the poet and activist Maya Angelou how she arrived at her commanding confidence onstage. "I know theater. I write for it. I have studied it. I act for it now and again. So I know theater. And I am centered," explained Angelou.

That's confidence based on evidence.

Shoot the thousand daily jump shots. Practice your instrument every evening. Apprentice under the best chef in town. Study the greats in your field. Keep putting yourself out there, gradually increasing the challenges you undertake. Pay close attention when the going gets tough. By working through current challenges, you are developing self-efficacy for future ones.

"When I left what was a very good job to write my first book, I didn't *believe* I could do it. That would have been absurd. What would that belief have been based on? I had never done it before," explains bestselling author Ryan Holiday. "What I did have was *evidence* of the traits necessary for success. I had put in the training as a research assistant on other books. I had written on a regular basis for many years (every day for six years in fact). I knew I wasn't a quitter. I knew I had mentors I could go to for advice when I felt in over my head."

If you are feeling unsure—be it on the start line of a marathon, the first page of an exam, or day one of launching a new business—the best thing you can do is pause to reflect on the evidence you've amassed.

If you've put in the proper training, you can have the faith to trust it.

Own Your Seat

Just because you have confidence does not mean that you'll never experience doubt.

A few years back, I was working on a piece for *Lion's Roar*, a popular magazine focused on ancient wisdom for modern times. In the essay, I kept quoting other thinkers to help make my argument. After the second round of revisions, the editor wrote me, "This is good, but I want to hear more about what YOU think. Not what Jack Kornfield or Tara Brach thinks, but what YOU think. Own your seat."

The last three words of her email—own your seat—have stuck with me ever since.

I was suffering a classic case of imposter syndrome. By this point in my career, I had published two books and written for *The New York Times* and *The Wall Street Journal*. It didn't matter. This was a new publication, a *wisdom* publication. Who was I to offer wisdom? I didn't feel like my voice belonged.

When I was sharing my experience with a friend, she reminded me that anyone who says—and worse yet, genuinely believes—they have it all figured out is a person to run away from, and fast. That I was hesitant about writing the piece wasn't a bad thing. If anything, it was a good thing. The problem wasn't my lack of wisdom; it was that I was comparing myself to some illusory bar of enlightenment. That I felt a bit uncertain was completely normal—after all, there are few things in life about which we *should* feel certain.

Owning my seat didn't imply having 100 percent assurance. It meant realizing I'd done plenty of thinking, research, writing, and coaching on the piece's topic, "right striving." Perhaps I didn't have it all figured out, but nobody does. Even so, I could be confident enough because I'd done the work.

A good editor, coach, manager, mentor, or leader won't call you up to bat until you are ready. It does not mean the at-bat will be comfortable, nor does it guarantee success. But you can own your seat nonetheless, stepping into the arena, taking your doubts along for the ride, and relying on evidence to remind yourself you belong.

In the end, I wrote the piece. The best part wasn't being published in *Lion's Roar* or the $300 I got paid. It was the opportunity to wrestle with the idea of owning my seat and to explore the interplay of confidence and doubt.

Nowadays I frequently ask myself, *Do you have what you need to own your seat? If so, own it. If not, what evidence do you need before you can?*

These are good questions to ask yourself, too.

Humility

In the same 1987 interview with Mavis Nicholson, Maya Angelou explained, "I hope to be humble. Humility is from within. And that keeps you honest."

Angelou's magnetic presence owed itself to the combination of her confidence and humility. These qualities run together. When we have confidence, it means we've done the work to gain it. Doing the work to gain it is hard. And doing hard things makes us humble. It teaches us that we can push limits, but also that we have limits to begin with.

Humility doesn't just lead to charisma. It is also leads to growth.

Once we think we know everything, we cease to keep knowing. Once we think we are the best, we cease to keep getting better. But there is always more to know, always room to improve. Knowledge and skill are continually evolving and advancing. If we want to evolve and advance with them, we must maintain an open mind. This is easier to do when we have confidence. Someone who is insecure perceives everything that challenges them or their views as a threat. Someone who is confident welcomes challenges, knowing at least some will make them better.

There are layers and layers to craft and expertise. A big part of the thrill and satisfaction of excellence comes from continual exploration and deepening our relationship with an activity. Those who speak softly, tread lightly, and embrace the words *it depends* tend to possess the most confidence. A bit of doubt is okay—it's often a sign that we are exactly where we need to be.

Social scientists call this *intellectual humility,* which can be understood as owning your seat, and, at the same time, realizing that nobody is infallible. It means not worrying about always being right or needing to exert power or control over others to feel good about oneself. Research shows intellectual humility is associated with enhanced self-awareness, curiosity, and discernment. It makes us better at whatever we do.

When we earn confidence, we also gain humility. We learn to own our seat while being open to growth and development. We inch toward the magnetic presence exuded by someone like Maya Angelou. When people talk about being in the presence of greatness, this is what they mean.

Chapter Summary

- The faith required for excellence is not blind or delusional; it is acquired.
- Confidence comes from evidence.
- You gain self-efficacy by working through challenges.
- Own your seat.
- Confidence and humility go hand in hand.
- When you have confidence, you don't perceive everything that challenges you as a threat. If anything, you welcome challenges, knowing at least some of them will make you better.
- Confidence is not the absence of doubt; it is facing doubt with empirical support and stepping into the arena no less.

Confidence comes from evidence.

Trust your training.

12

Patience

From the time he first had the idea of evolution, it took Charles Darwin more than twenty years to publish his masterwork, *On the Origin of Species*. That's about the same number of years Shalane Flanagan had been running competitively before she finally won the New York City Marathon, solidifying her as one of America's greatest distance runners. Steve Jobs had been at Apple for thirty years prior to the first iPhone being released. Katalin Karikó tirelessly researched mRNA for nearly forty years before the development of lifesaving vaccines reliant on her technology.

These stories are not outliers.

A 2018 study published in the journal *Nature* examined peak performances in creative and intellectual pursuits and found they all had one thing in common: Each rested on a foundation of prior work, during which observable improvement was less substantial. If the individuals mentioned above had given up or switched approaches prematurely, their advances never would have occurred. The world would be a very different place.

The martial artist and actor Bruce Lee once said that "patience is not passive, it is concentrated strength."

The media loves to cover the stone cracking, but what we don't see is the years upon years of tension leading up to that point. In his personal diary, Darwin wrote that he owed his discovery to "the love of science and unbounded patience in long-reflecting over any subject."

There is no such thing as an overnight breakthrough.

Patience is one of the greatest competitive advantages there is. It neutralizes our inclination to hurry and overemphasize acute results, and it lends itself to strength, stability, and lasting progress.

I've yet to meet someone who has achieved excellence and who describes their process as being rushed.

A Culture Addicted to Speed

Societally we've gradually been shifting away from the practice of patience. A gap in time becomes an opportunity to check notifications or scroll social media; food is ordered with the click of a button; short-form, sensationalized content is engaged with more than its long-form counterpart. We want instant health, instant wealth, and instant happiness.

In 2006, the research firm Forrester found that online shoppers expected web pages to load in under four seconds. Three years later, that number was compressed to two seconds. By 2012, Google engineers learned that internet users expect search results to load within a mere two-fifths of a second, about how long it takes to blink. There's no reason to believe this trend is slowing down.

The author Nicholas Carr, whose work explores the far-reaching effects of the internet, says that "as our technologies increase the intensity of stimulation and the flow of new things, we adapt to that pace. We become less patient. When moments without stimulation arise, we start to feel panicked and don't know what to do with them, because we've trained ourselves to expect this stimulation."

A report conducted by the Pew Research Center's Internet and American Life Project supports Carr's assertion. After reviewing survey data from a large sample of internet users, they concluded a likely side effect of our hyperconnected world is an "expectation for instant gratification."

There's nothing inherently wrong with expedient technology—I rely on it, and I'm just as likely as the next person to become frustrated when a web page is slow to load. But when we expect this kind of speed, stimulation, and instant gratification in all areas of our lives, it becomes problematic. That's because to make a meaningful difference in just about anything of consequence, the work you put in needs to persist long enough to break through inevitable barriers and plateaus. What seems like a static period may not be a static period at all; you may just not yet be seeing the effects of your efforts.

Timothy Wilson is a social psychologist at the University of Virginia in Charlottesville. Like Carr, he had a hunch that people were becoming increasingly addicted to speed and novelty and incapable of handling waiting and boredom. To test his hypothesis, Wilson recruited hundreds of undergraduate students and community members to partake in what he told them were "thinking periods." Participants were placed in empty rooms for fifteen minutes without anything to distract them. Their smartphones, laptops, and notebooks were confiscated. Wilson gave the participants two options: sit and wait out the fifteen minutes or shock themselves with a strong electrical current.

Sixty-seven percent of men and 25 percent of women chose to shock themselves, often repeatedly, rather than sit still and wait. These people weren't self-selected gluttons for discomfort. Before the study, all the participants said that they would pay money to avoid shocking themselves. Yet when it came to sitting and waiting—again, for a mere fifteen minutes—a majority of men and a sizeable proportion of women preferred putting themselves through physical pain.

Wilson's study may be an extreme example, but we do the equivalent of shocking ourselves instead of practicing patience all the time. It's when we rush our process only to end up injured or burnt out; when we take so-called shortcuts only to end up back at square one; or when we prematurely give up on something we believe in.

Excellence is a long and windy road that is full of ups, downs, and everything in between, including periods of boredom. As such, it requires a steady pace and patience. It requires getting comfortable with the discomfort that Wilson's study participants could not bear, and on a much longer horizon.

"To take the master's journey," writes the Aikido master George Leonard, "you have to practice diligently, striving to hone your skills, to attain new levels of competence. But while doing so—and this is the inexorable fact of the journey—you also have to be willing to spend most of your time on a plateau, to keep practicing even when you seem to be getting nowhere."

Patience does not come naturally. Like any skill, it must be developed. One small but powerful way is to leave your phone or other digital device behind when you go about the usual activities of your day. Consider keeping your phone in the car when you go into a grocery store or run errands. If you end up having to wait in a checkout line for a few minutes, that's great—you'll get practice enduring it. Another opportunity to practice patience is when you are out to dinner with someone and they go to the restroom. See if you can't sit and wait with yourself instead of immediately pulling out your phone.

Not having our phones during short periods of waiting may seem trivial, but it helps decondition our habituation to novelty and speed. This carries over into larger areas of life. If we can unwind our dependence on stimulation, if we can get comfortable with open and liminal spaces, then we can make more intentional decisions about when to move fast and shake things up versus when to slow down and stay the course.

Stick-to-itiveness

There's a popular saying that people overestimate what they can accomplish in a year but underestimate what they can accomplish in a decade. Research backs it up.

Studies show the average age of a Nobel Prize–winning scientist when they conduct their groundbreaking work is forty-four years old. Katalin Karikó, the mRNA researcher mentioned in the opening of this chapter, was forty-two when she joined the lab that supercharged her science.

A similar pattern is true in business. For a massive project examining the success and failure of start-ups, economists from Northwestern and MIT examined all businesses launched in the US between 2007 and 2014. The dataset encompassed 2.7 million founders. The researchers compared a founder's age to a variety of company performance measures, such as employment, sales growth, and, when relevant, a company's value at initial public offering (IPO). They found that successful entrepreneurs are more likely to be middle-aged than young, and are more likely to experience success after not one year in the game but a decade and in many cases more.

For the top 0.1 percent of the fastest-growing businesses in America, the average age of the founder when their company was first launched is forty-five. Middle-aged founders also have the most successful IPOs. A fifty-year-old founder is 1.8 times more likely than a thirty-year-old founder to create a high-growth business. Even founders who start their companies when they are young may not peak until later in life, according to the researchers.

"These findings strongly reject common hypotheses that emphasize youth as a key trait of successful entrepreneurs," write the study's authors. "The view that young people produce the highest-growth companies is in part a rejection of the role of experience."

Imagine two curves: We'll call the first "raw ability" and the second "wisdom." For most pursuits, raw ability rises and peaks between the ages of sixteen and thirty. The wide range accounts for the fact that a gymnast or football player is at their best raw ability in their late teens or early twenties, whereas someone whose craft is less reliant on physical attributes, such as an investor or scientist, may not see their raw ability peak until a few years later.* The wisdom curve, on the other hand, starts at zero and gradually rises as you gain more experience in whatever it is you do. When these

* Researchers call this *fluid intelligence*, or thinking quickly on your feet. Studies show, on average, it peaks anywhere between twenty and thirty.

two curves—raw ability and wisdom—intersect, you are primed for your best performance.

The slope of the curves varies by task. For example, in sports that rely heavily on physiological fitness, such as sprinting one hundred meters, the decline of the raw-ability curve would be steeper than in a sport such as alpine climbing or orienteering, where age-related fitness declines are slower and less significant. The same could be said for the difference between intensive financial modeling and writing poetry. The former relies more on raw ability, and the latter relies more on wisdom.

Regardless of our activity, we need to stay in the game long enough to gain at least some wisdom before we can achieve our best. This often takes years or even decades. It explains why scientists and founders peak around forty-five and not twenty-five, and why athletes generally spend years competing before they achieve their top performances. It's imperative to zoom out and adopt a mindset that measures success not in weeks and months but in years and decades.

It's not to say you can't be great at twenty-six or that all hope is lost after your forties. Far from it. First off, these numbers are averages. Some people peak early, and others peak late. Whatever age you are right now, you should give your best effort, knowing the deposits you make will compound in the future. Roger Federer had the highest-winning percentage of his career at age thirty-seven. Many artists, knowledge workers, and craftspeople do their greatest work well into their fifties, sixties, and seventies.

Second, as you age, your contributions and insights may shift toward wisdom and mentorship. These qualities are as vital as anything, and the world desperately needs more of both. The poet Mary Oliver published some of her most incisive verses in the final decade of her life. Though he was no longer seeing patients or making diagnoses, the neurologist Oliver Sacks penned a series of poignant medical essays from his deathbed. The work itself may evolve, but the drive underneath it remains.

Pop culture says progress should be fast, predictable, and easy. The truth is that progress is often slow, nonlinear, and more challenging than you think. The actual secret to greatness:

> Pick your thing.
> Pick a good system for your thing.
> Surround yourself with people who support you doing your thing.
> Do your thing for a decade (or more).

Little Victories

Winning an Olympic medal might take fifteen years, but you can focus on executing your daily workouts. Publishing a bestselling book might take twenty, thirty, or even forty years, but you can write one thousand words you are proud of every week. The same goes for launching a business, creating art, doing science, and so much more.

Embracing a process mindset is paramount. A process mindset creates opportunities for little victories, which help us to sustain the motivation and stick-to-itiveness required to accomplish big goals. The master, wrote the Taoist philosopher Lao-tzu, "accomplishes the great task by a series of small acts."

A handful of studies provide insight into why this is the case. Researchers have found that when we accomplish micro-objectives on the path to distant goals, our bodies release dopamine, the neurochemical associated with motivation and drive. Without hits of dopamine, we are likely to become apathetic and give up. In other words, process promotes progress, and progress, on a neurochemical level, primes us to persist.

Breaking goals down into their component parts and then focusing on those parts is essential to patience, and therefore it is essential to excellence.

Katalin Karikó says her steadfast determination comes from her love for "the bench," or the spot in the lab where she works. Though Karikó's major breakthrough took decades, it was also the culmination of hundreds of smaller milestones, many of which she and her colleagues defined for themselves. When asked about her newfound fame in a recent interview, she replied, "The bench is there. The science is good. Who cares?" These are the words of someone who lives for the process, and whose patience comes naturally as a by-product.

It's not to say patience is easy. As Wilson's study showed, waiting can be painful. When we are working hard toward a goal and want it badly, there is a propensity to zoom in and lose perspective. It happens to me, and it probably happens to you, too. But if we can stay with the discomfort of waiting, if we can just hold on and avoid making reckless decisions and big mistakes, then we improve our odds of accomplishing something extraordinary. Stick to the process. Focus on the work itself. You can't control what may or may not be down the road. All you can control is your effort right now.

Increase Your Surface Area of Luck

Karikó's persistence allowed her to stay in the game long enough to get lucky.

To be clear, she worked as hard as any scientist in her generation.

But that's not to say fate didn't play a role. The dichotomy between hard work and luck is a false one. All great accomplishments include some measure of both. And while you can't control luck, you can improve your chances of it happening.

The mRNA technology Karikó developed is particularly effective for viruses that feature a "spike protein" on their surface. It just so happened that one such virus, COVID-19, took the world by storm in late 2019. In order to be turned into an effective vaccine, though, mRNA needs to be encased in a microscopic amount of fat so that it can enter cells. Fortunately, a team of researchers in British Columbia had been working on creating that exact delivery mechanism, called a lipid bubble, for more than twenty-five years. Karikó's historic breakthrough was the result of these variables coming together at the right time.

She couldn't have predicted the emergence of a new virus that would be a perfect fit for her mRNA technology. Nor could she have predicted another team of scientists would be finishing their labor on a complementary technology that was necessary to make everything work. Those parts were luck. But by staying in the game, by continuing to put herself and her work out there, Karikó and her research partner, Drew Weissman, increased their surface area for luck. Now, their mRNA technology is showing great promise to treat an array of diseases, including many types of cancer. Time will tell if that promise comes to fruition.

Luck plays a role in everything. You never know when the weather will be right. Or when the market will be right. Or when the competition will be right. Though we may not like it, many factors are outside of our control. What we can control, however, is staying in the game and continuing to show up. We simply cannot know when the universe will conspire to gift us the latticework of variables that will bring out our best. All we can do is remain patient and put ourselves in a position to receive it.

Magnus Carlsen, the highest-rated chess player in history, doesn't lose often, particularly in classical games, a format of play in which time pressure is removed. However, at the 2025 Norway Chess tournament, he made an uncharacteristic mistake. His opponent, nineteen-year-old Gukesh Dommaraju, jumped on the error, and preceded to beat Carlsen a few moments later, in a game that lasted sixty-two moves.

"Ninety-nine out of one hundred times I would lose," Gukesh said in an interview following the match. "Just a lucky day."

Carlsen had been in a commanding lead. He had outplayed Gukesh for much of the game, establishing a near winning advantage. But instead of resigning or phoning it in, Gukesh held tough. You never know what is going to happen. That's why you play the game. That's why you give yourself a chance. Gukesh may only win that match against Carlsen one out of one hundred times, which is why it's so important to show up for the other ninety-nine.

Consider a series of studies that examined the connection between the quality of one's work and the quantity of one's work and found that the latter drives the former. "Quality is a probabilistic function of quantity," says one such study's author, the psychologist Dean Keith Simonton. In other words, the more we produce, the better we become. Vincent van Gogh created two of his most famous paintings, *The Starry Night* and *Sunflowers*, just two years before his death, following *hundreds* of other paintings.

The relationship between quantity and quality can be explained by two primary causes: The first traces itself back to luck; if we keep showing up and producing, we improve the odds that eventually everything comes together. The second is that the longer we do something—the more reps we put in, the more we experiment and figure out what works and what doesn't—the more we improve.

None of this is to say we should never quit. Throwing yourself against the same wall only to repeatedly fall down is not a particularly wise strategy. If Gukesh had no chance of winning his match against Carlsen, or if the costs of playing on would have been detrimental, then continuing would have been foolish. But if you have a path to victory and you are enjoying the process itself, then there are vast benefits of staying in the game, even, and perhaps especially, when it's hard.

In the words of the Iowa State football coach Matt Campbell, "If you fall in love with the process, then eventually the process loves you back. But see, here's what's crazy about that. You don't know when it is going to love you back. All you have to do is be prepared for your opportunity when it's ready to love you back." This sentiment is true for athletes, but it's equally true for creatives, researchers, artists, and really for all of us. No artificial intelligence, algorithm, or automated process can predict when a breakthrough is going to occur. It's why we describe the experience as being *magical* when it happens. That unpredictable magic, and the incredible feelings of aliveness and accomplishment that accompany it, is a big part of what keeps us coming back to our pursuits. It's a big part of what makes excellence so exhilarating.

Most things that are worthwhile take time. As they say in baseball: Be patient, it's a nine-inning game.

Though we long to be in the bottom of the ninth with a seven-run lead, the truth is, we may still be in the middle of the third. Excellence is a long-term endeavor, and some innings are better than others. In numerous situations, both personal and professional, we convince ourselves the current inning is the only one there is and ever will be. But if we zoom out, we realize so many of the meaningful projects in our lives are nine-inning games. The perceived immediacy of whatever it is we are dealing with relaxes, and we grow to realize the fundamental value of patience and persistence.

Excellence is not about being a champion of one inning.

It's about playing all nine as best we can.

Chapter Summary

- There is no such thing as an overnight breakthrough.
- Our work must persist long enough to overcome inevitable barriers and plateaus.
- Deconditioning ourselves from speed and novelty is a proactive practice.
- When everyone is about quick fixes, hacks, and fads, be about stick-to-itiveness: It is essential to excellence.
- Accomplishing micro-objectives en route to larger goals primes us to persist.
- If we love the process, eventually the process loves us back.
- The longer we stay in the game, the more our surface area for luck increases.
- Keep producing: Quantity helps *create* quality.
- Excellence is a nine-inning game.

The actual secret to greatness:

Pick your thing.

Pick a good system for your thing.

Surround yourself with people who support you doing your thing.

Do your thing for a decade (or more).

13

Routine

Some people have elaborate, optimized morning routines. Other people, like myself, hit snooze on the alarm, stumble to the coffee maker, and putter about as they gradually come to life. There's been a deluge of morning-routine content on social media and in the pages of magazines, newspapers, and books. According to Google's Ngram Viewer, which tracks the appearance of words and phrases in books, "morning routine" has nearly quadrupled in the last two decades. If you don't follow an intricate series of steps upon rising, it can make you feel like you are doing something wrong. But the truth is more complicated.

Every day we wake, we face a messy and complex world rife with uncertainty. We cannot control the weather. We cannot control other people. We cannot control the feelings we experience from moment to moment. If you believe otherwise, try for an hour and see how it goes. Yet there are things we can do to create order amid the chaos and settle into a rhythm. Herein lies the power of routines.

Nearly everyone who attains excellence relies on routines. Routines provide structure to our days, offer a sense of predictability in an inherently unpredictable world, help us to activate when we are feeling low, and keep us grounded when we are bursting with energy. They automate decisions so we don't burn willpower and prime our minds and bodies to perform. Routines support focus, consistency, discipline, patience, confidence, and many other factors of excellence. If you work out, write, sketch, or journal every morning, you don't have to think about it. You just do it. It becomes part of your routine.

Research also shows the objects with which we surround ourselves elicit certain behaviors. For example, the more you pair going to a specific coffee shop at a specific time of day with writing, the easier it becomes to enter into a productive flow. It's why people find comfort in having a game-day blazer for public speaking, or a particular pair of jeans they wear to the artist's studio. These artifacts serve as a signal to step outside of ordinary life and into your respective craft. The best routines are like easy chairs, writes George Leonard in *Mastery*. You settle into them, unaware of the time and turbulence of the world.

Here's the catch: Although routines can be magical, there is no magic routine.

Many features found in so-called "optimal routines" affect people differently. Some perform better while listening to music. Others do not. Some get a boost from caffeine. Others experience anxiety or an upset stomach. Some find a cold shower energizing. Others shiver for hours and find the whole ordeal exhausting.

We cannot develop optimal routines by mimicking what other

people do. While it's true that certain behaviors are close to universally effective—such as exercise, social connection, and sleep—even then, there is no single best time, place, or way to engage in each. It is only through astute self-awareness and experimentation that we discover what works best for us. The remainder of this chapter will show you how.

Different people have different chronotypes, a term used to describe the natural ebb and flow of energy we experience over the course of twenty-four hours. Whether it's a physically or cognitively demanding task, research shows most people tend to perform at their best in either the earlier or later part of the day. These individual differences are rooted in our unique biological rhythms—when various hormones associated with energy and focus are released and when our body temperature rises and falls. Scientists refer to those who are most alert in the morning as larks, and those who are most alert in the evening as owls. There is no evidence that either chronotype is inherently better. What matters for each of us is trying to align our activities with our energy levels.

In his book *Daily Rituals,* the author Mason Currey detailed a typical day for over fifty of the world's greatest artists, writers, musicians, and thinkers. Nearly all of them had tried-and-true routines. But the routines themselves varied significantly. This was especially true for *when* people did their best work. Some, like Mozart, worked late into the night. Others, like Beethoven, were most productive at the crack of dawn. The take-home message wasn't that the majority of great performers accomplished their best work at a certain time of day or that there is an optimal hour for productivity. Rather, each individual figured out when they were most alert and focused, and did what they could to design their days accordingly.

Designing Days

Researchers from the Sleep Research Center at Loughborough University in the United Kingdom developed an evidence-based questionnaire to help people determine their chronotype. Answering the following three questions gives you a good idea of where on the lark-owl spectrum you fall:

1. "If you were entirely free to set up your evening, with no commitments in the morning, what time would you go to sleep?"
2. "You have to do two hours of physically hard work. When would you do this work if you were entirely free to plan your day?"
3. "You have to take a two-hour test, which you know will be mentally exhausting. When would you choose to take the test if you were free to choose?"

This questionnaire is a valuable tool, especially when paired with listening to your body in real time. For the next few days, you could pay attention to when your energy levels are highest and also when you fall into a foggier brain state where your attention lags and your work starts to suffer. Most people recognize their chronotype after just a day or two of paying close attention. However, the gold standard is to go seven days without setting an alarm clock or compensating for fatigue with caffeine or other stimulants. Not only will you accurately zero in on your chronotype, but you'll also benefit from a reset period during which your body can return to its natural rhythm.

Once you have a sense of your chronotype, you can take advantage of it by intentionally scheduling your hardest and most demanding work for when your alertness peaks. When your biology shifts and your alertness diminishes, you could focus on tasks that demand less attention, such as responding to emails, scheduling unavoidable meetings, or doing basic chores around the house. Finally, when your attention begins to wane completely, you need not force yourself to keep going. Rather, you can let your mind and body recover.

To be sure, there are as many unique situations as there are people in the world. Family, colleagues, teammates, and other obligations all place constraints on what you can and can't do, and also when. But even having just a general sense of your chronotype running in the background is advantageous. For instance, I've come to learn that I do my best work in the first half of the day, and that my brain is essentially useless after 6:30 p.m. Does that mean I write uninterrupted every day from six in the morning to noon, or that I never do any work in the evening? No. But I get much closer to that than I would if I didn't keep my chronotype in mind. The goal is not to be rigid or perfect. The goal is to use this information as best you can. Instead of thinking about managing your time, think about managing your energy.

Daily, Weekly, and Monthly Practices

Another powerful way to think about routines is to develop daily, weekly, and monthly practices. These become foundational elements of our lives that prime us for excellence. Rather than trying to concoct elaborate routines with nineteen components—and in the process, generating more stress rather than alleviating it—we can simplify and focus on what matters most. We can name the main things and keep them the main things.

My three daily practices are forty-five to ninety minutes of physical activity, at least one block of deep-focus work on a meaningful project, and not fighting evening sleepiness, which usually means going to bed before 10:00 p.m.

My three weekly practices are at least two long walks outdoors, gathering with friends at least once, and spending one day completely offline (a digital sabbath).

My three monthly practices are at least one chunk of meditation, contemplation, listening to music, or some other form of deliberate reflection on who I am and why I'm here; at least one day when I'm predominantly outdoors; and at least one outing when I'm involved with my local community.

These practices and rhythms work for me. Yours will almost certainly be different. But the broad catalog—movement, sleep, community, nature, meaning—is aligned with decades of research on health, longevity, and flourishing.

The other area to explore is nutrition. What we put into our bodies can have a profound impact on how we feel and what we do. Whereas some people find rules around food unhelpful (and in the case of disordered eating, quite harmful), other people find them beneficial. Examples include eating a certain number of vegetables daily, avoiding fast food, limiting dessert to special occasions, or developing a general menu of options for breakfast, lunch, and dinner.

It cannot be stressed enough that there is considerable variability from one person to another. We can be confident that coming up with daily, monthly, and weekly practices is beneficial, but only you can figure out what those practices ought to be.

What makes an approach like this powerful is its simplicity and concreteness. Developing a few anchors fosters stability, predictability, and strength in a world of chaos, complexity, and infinite choices. I've yet to meet someone who hasn't found this framework effective, if not transformative. Remember that it is the seemingly small day-to-day routines that set the foundation for peak moments and even greatness. Everyone loves to talk about the latter, but we don't get there without the former.

The simpler our lives, the more stringent we can be with our routines. Someone who lives alone and sets their own working hours may find it advantageous to do most things at a certain time and place. Someone with children and a variable profession will almost certainly need a broader and more flexible set of principles. Neither approach is better or worse. Problems only arise when you try to fit a square peg into a round hole. The point is that routines are personal. What works for one person may not work for another. What matters is being cognizant of the trade-offs and then intentionally choosing the best routine for you.

Transitions

Scott Kelly is a former NASA astronaut who, in 2016, spent 340 days on the International Space Station (ISS), the longest time an American has ever been in space. His experience is akin to not leaving your house for nearly a year. In a CNN interview that focused on working from home, Kelly emphasized having a set schedule: "If you're lucky enough to be able to work from home, you know, schedule those work times. I would go as far as even scheduling meals. My wife and I have been making a schedule like we were in space, because if you keep to that schedule, and it has variety, I think what people will find are the days go by much quicker."

Kelly is alluding to the importance of developing markers for time and place. Before the digital revolution, many of these markers were automatically built into our lives. Most of us would drive to and from work, and when we left work, we'd really *leave* work. But ever since the introduction of laptops and mobile devices, work and life more easily bleed into one another. While the ability to work from home comes with significant benefits, an overlooked cost is the disappearance of transitions between the different parts of our lives. Without transitions, one day merges into the next, time feels hazy, and we never really disconnect.

Whereas transitional routines were once automatic, now many of us must create our own. They need not be elaborate. Examples include reading for fifteen minutes before starting the workday, making an afternoon pot of coffee or tea, or marking the end of the day with a brief walk outdoors. The particulars don't matter; what matters is regularly doing something to distinguish the different segments of our days.

The Three Rules of Routines

Despite their vast benefits, becoming overly reliant on routines can be hazardous. If you can't stick to your routine for whatever reason—you're traveling, your favorite coffee shop closes, you enter into a new relationship or have children—you won't know what to do.

You want your adherence to routines to be rugged and flexible. If you are not rugged, you'll get pulled this way and that by the whims of life, and it will be near impossible to show up at your best consistently. But if you are not flexible, if you try to rigidly stick to your routines in all circumstances and at all costs, you'll build up tension and stress until one day you break.

1. The first rule of routines is to develop them and do everything you can to uphold them.
2. The second rule of routines is to release from them when life demands it.
3. The third rule of routines is not to let rule 2 become a chronic excuse for rule 1.

Chapter Summary

- Routines provide structure and stability in an inherently messy and turbulent world.
- There is no such thing as a magical routine. You've got to experiment and figure out what works for you.
- Do what you can to design your days in alignment with your chronotype.
- A simple yet powerful framework to lay the foundations for

excellence: three daily practices, three weekly practices, and three monthly practices.

- Mark transitions between the different spheres of your life; this helps prevent the feeling of everything blurring together and time collapsing on itself.
- The best routines are rugged and flexible.

It's the seemingly small day-to-day routines that set the foundation for greatness.

Everyone loves to talk about the latter, but you don't get there without the former.

14

Gumption

In his decades of writing and repairing motorcycles, Robert M. Pirsig came to believe that nothing is more important than gumption. "It's the psychic gasoline that keeps the whole thing going," he writes in *Zen and the Art of Motorcycle Maintenance*. "A reservoir of good spirits."

Gumption is a forward inertia, a sense of progress and possibility, a strong yet measured enthusiasm. As you'll see in the coming pages gumption is not an innate trait; with the right mindset and actions, nearly anyone can develop it. When you have gumption you perceive obstacles as challenges rather than threats; you problem-solve and bring your creativity to bear. Gumption doesn't come to the fore in the absence of obstacles; rather, it's a feeling that no obstacle can stop you.

"A person filled with gumption doesn't sit around dissipating and stewing about things. He's at the front of the train of his own awareness, watching to see what's up the track and meeting it when it comes," explains Pirsig. "If you haven't got it there's no way the motorcycle can possibly be fixed. But if you *have* got it and know how to keep it there's absolutely no way in this whole world the motorcycle can *keep* from getting fixed."

Gumption emerges out of a commitment to caring, consistency, focus, discipline, patience, and all the other factors we've discussed so far. When we're filled with gumption, we have a general sense of where we're going and how we'll get there, but we also pick up on real-time cues, making micro-adjustments along the way. It's when skill, attentiveness, and energy overlap; when we enter into an intimate and engaging dance with our pursuit, be it training for a marathon, molding pottery, or running a restaurant. We sense what is happening around us and respond at the height of our powers. We gain a feel for what we're doing. We settle into a groove.

And yet, even the most productive paths of excellence include difficulties and downturns. There are missteps, illnesses, injuries, periods of boredom, comparisons to others, and overly critical self-judgments. Pirsig calls these *gumption traps*: "They drain off gumption, destroy enthusiasm, and leave you so discouraged you want to forget the whole business," he explains.

Gumption traps come in two categories. The first is external, or when things outside of yourself interrupt your rhythm and deplete your energy. Examples include the weather, competition, actions and opinions of others, and shifts in cultural values and vibes.

The second category is internal, which is when you get in your own way. Examples include rigidity, ego inflation, doubt, and overwork.

With external gumption traps, it feels as if the universe is conspiring against you. With internal gumption traps, it feels as if you are conspiring against yourself.

There are as many gumption traps as there are people and pursuits. Although it is impossible to predict which traps you'll face, a few strategies can help you build, maintain, and when necessary, regain gumption in nearly any pursuit. We'll turn to them now.

Come Back to the Work

Do something for long enough, and you are going to experience frustrations and failures, successes and achievements. You want to learn from the former and relish the latter but also not get too caught up in either, lest you find yourself on an unsustainable emotional roller coaster. It can be helpful to give yourself a set amount of time to grieve defeats and celebrate victories, but after that, get back to doing the work itself. I've come to call this *the forty-eight-hour rule*: give yourself that much time to sit with victory or defeat, but then return to the work. Of course, the exact timing is flexible. For bigger wins and losses, perhaps you'll need a week or two. For smaller ones, maybe you can recover in an afternoon. What matters is that you set a cap on dwelling over the result and get back to the process.

Doing the work has a special way of putting both success and failure in their respective places. The work doesn't change nearly as fast as our emotions. Win or lose, great day or terrible day, the blank page is still the blank page, a lap in the pool is still twenty-five meters, and the ideas on the whiteboard continue to benefit from organization.

Returning to the work keeps our focus, joy, and meaning rooted in the process, not the outcome. It reminds us of why we got into our activities in the first place. It ensures we don't become overly attached to success or failure, each of which comes with its own trappings, be it addiction to external validation, complacency, or self-doubt.

After the swimmer Katie Ledecky wins gold at the Olympics and all the other competitors have filed out of the arena, she remains alone in the pool working on her stroke, making subtle adjustments, trying to get better even though she's just been crowned the best. The legendary songwriter Bob Dylan is known to make brief appearances at award shows, if he shows up at all. He doesn't care about the awards; what he cares about is the work. Ledecky has won ten Olympic medals. Dylan has released fifty-five major albums and won ten Grammys.

These examples remind us to return to the work, and generally sooner than we think. If we get too caught up in the results, good or bad, it becomes harder to sustain gumption for the process. But it's within the process that we find lasting fulfillment, enthusiasm, and energy.

See the Ball Go Through the Net

When a player is slumping in basketball—whether it's missing shots he normally makes or turning the ball over and becoming frustrated—the best thing is to take a layup or get to the free throw line. Both are high-percentage shots for most skilled players. When you make a layup or free throw, you put points on the board and see the ball go through the net. You regain a little confidence and feel for the game, which makes it easier to leave the slump behind.

Most activities have the equivalent of a layup or free throw.

At a logjam drafting your book's manuscript? Write a newsletter, email, or blog post. Feeling powerless as a leader during a challenging situation? Walk the halls of your workplace or make a few phone calls to engage with individual colleagues. Struggling to execute a new recipe and spiraling into discouragement as a result? Return to an old favorite you know like the back of your hand. Falling short in all of your workouts? Ditch the watch or timer and complete one of your favorite sessions by feel.

The point is that when you find yourself in a rut, stacking up a few victories, no matter how small or basic, goes a long way to help.

Step Away

When I find myself stuck on a physical challenge or creative project, my default reaction is to work through it—to try to bust the plateau or force a creative insight.

Sometimes this works. Often it doesn't. In the latter instance, it is almost always advantageous for me to step away. But that doesn't make it easy.

The human brain hates leaving things undone. It's why we face a strong urge to lean in and continue picking at problems even when we are getting nowhere. Yet research from across fields shows that we get unstuck *not* when we are working on something but rather when we step away. This is precisely why we experience aha moments in the shower or while walking or driving home from work. It is also why some of the best physical feats happen following what seems like a too-long break. After multiple failed attempts, Roger Bannister finally became the first person ever to run a mile in under four minutes—but not until he took a hiatus from training that lasted two weeks, which he spent hiking with friends.

The million-dollar question is *when* to step back.

The last thing you want is to prematurely shut things down when a breakthrough is around the corner. Two ideas can help. The first is when it feels like you are straining hard on something that, when you are at your best, comes naturally. The second is when an activity you normally find interesting and exciting begins to feel boring or tedious. Both are signals that you may benefit from stepping away.

Sometimes we need to keep grinding in order to break through, but these instances are exceptions that prove the rule. What usually happens is that the more we lean in, the worse we feel and perform. We stay up working until midnight, only to redo everything the next morning. We keep pushing through a workout, only to aggravate an old injury. We're almost always better off with the extra sleep or taking a rest day. If we find ourselves losing gumption and react by becoming desperate and trying to cram in more work, we set ourselves even further back. It's a tale as old as time.

The good news is that if you pay attention you'll begin to pick up on patterns, realizing subtle signs that tell you when to push forward versus when to step away. For instance, when I catch myself making circular edits—changing a sentence one way, then on the next read changing it back—that's a sign I am ready for a break. Sometimes that break is an hour, a day, or even a week. Upon returning, I discover a refreshed enthusiasm for what I found tiresome and tedious. My writing improves.

The Grammy award–winning violinist Hilary Hahn told me that a big part of her practice is knowing when to lean in and "grind," as she puts it, versus when to step away. A huge unlock, she says, was learning "not to view stepping away or taking a day or two off as antithetical to my consistency but rather to see stepping away at the right time as a crucial component of my consistency, of staying in an overall rhythm and maintaining momentum." By paying close attention to how she feels and what happens as a result, Hahn has learned when to stay with discomfort and keep pushing versus when to step back. "The creative process ebbs and flows," she says. "I've learned I need to ebb and flow with it."

Explore Outside of Your Domain

The psychologist Mihaly Csikszentmihalyi spent decades studying people who made major contributions in the sciences, arts, business, and government. He found that most of them "were inspired by a tension in their domain that became obvious when looked at from the perspective of another domain. . . . Some of the most creative breakthroughs occur when an idea that works well in one domain gets grafted to another and revitalizes it."

Marie Curie's physics informed her chemistry and her chemistry informed her physics. She won Nobel Prizes in both fields. Leonardo da Vinci was a polymath. His interest in science and anatomy heavily influenced his masterwork drawing *Vitruvian Man*.

It is especially helpful to step outside of our respective domains when we're feeling stale, stuck, or uninspired. A reliable way for me to renew gumption toward writing is by listening to music. I've met many musicians who say they renew gumption by reading books. We never know what we'll find in other fields that can provide new ideas and vigor for our own.

Even when we aren't stuck, we can stay fresh by spending time outside of our primary pursuits. A simple yet powerful method is reading across subjects and genres. There is no better place than a book to get a rich distillation of insights and wisdom. I've helped four-time Olympians overcome psychological bottlenecks with a single book recommendation. I've helped physicians and founders navigate rough waters the same way. From Abraham Lincoln to Winston Churchill to Jane Austen, there is a long line of individuals who attribute their sustained gumption to voracious reading.

You can explore outside of your domain in numerous ways, so there's no point in being more prescriptive than I already have. Find methods that work for you; consider integrating this practice into your regular routine; and keep it in mind when you find yourself at an impasse.

Mind Your Ego

Success comes with its own hazards.

When you hit a hot streak, you may begin to think you are invincible and get lazy with your process. Measured effort, consistency, discipline, and all the other attributes that fueled your rise give way to comfort, complacency, and navel-gazing. As a result, the big mistake, the trap you ought to see coming but don't, is lurking just around the corner.

An inflated ego is like methamphetamine for gumption. It puts you on a wild ride with great highs, but the crash is ten times more severe and damaging.

The best way to keep your ego in check is to surround yourself with people who will call you out on your bullshit. These are the colleagues and friends who aren't scared to tell you the truth, even when you don't want to hear it. It's easy to find people who provide encouragement and say *yes* to everything. It's hard to find people who will tell you *no* and that you're out of your mind. Both have a role, but the latter are more valuable.

A good ego-check partner is someone who knows you and your work well, commits to telling you the truth, and maintains a watchful eye for ego bloat.

Let's use my line of work as an example. There are many writers who mean well, produce genuinely good ideas, and experience early success. But if their egos grow unchecked, the next thing you know they are chugging full speed ahead down the conspiracy-theory railroad line. Or they let the pursuit of growth, fame, and followers cannibalize the pursuit of good, values-driven work, suffering emptiness, apathy, and burnout as a result. I'm sure your field has its own variety of ego traps. They aren't unique to writing but an aspect of human nature itself.

I'm convinced one of the main reasons I haven't lost my mind is because of a pact I made with my closest collaborative partner. Early in our careers, we promised to notify each other at the earliest signs of ego inflation—and it's a promise we've kept. Being a professional nonfiction author increasingly requires an online presence. As such, we are always checking each other to ensure we aren't spending too much time on social media or letting it turn our brains into sawdust. I know plenty of musicians, artists, and athletes who have similar arrangements.

A human ego will expand to fill the space it's given. As such, our egos require constraints. Since a growing ego is incapable of constraining itself, it's crucial to surround ourselves with people who can help.

Stay Flexible

Another common gumption trap is rigidity. The things that work, work great—until they get in the way. We must be able to release even from our strengths when necessary.

Grinding toward a goal with focus and determination is a key element of success. That said, sometimes we glorify perseverance, sticking with an approach solely to showcase our toughness, when we'd be better off trying something new. The ability to push ourselves to the limit and beyond is a superpower—until what we really need is a break. Unrelenting self-discipline can take us to the top of our game. But without caution, it can also lead us down a steep slope of isolation, resentment, and burnout. Routines are key to excellence, but at some point, the universe asks us to release from them.

Strength without flexibility leads to rigidity. Flexibility without strength leads to instability. The superhero idiom that our greatest powers become our gravest weaknesses contains more than a kernel of truth.

Whenever we find ourselves feeling drained and exhausted, it's wise to ask if we've been over-indexing on a certain strength when it might be beneficial to turn toward its opposite. If we've been overdoing self-discipline, maybe what we need is more self-kindness. If we've been holding on to too much structure, maybe what we need is some freedom. If we've been grinding too hard, maybe what we need is a break. It's all right to double down on your innate temperament—so long as you occasionally remember to look the other way.

"When the practice gets tedious, when I'm hitting walls, I know I'm about to make progress," says the violinist Hilary Hahn. "Being stuck isn't a bad thing. You've just got to find your way out. And some really interesting stuff can happen when you do."

Hahn's experience is not unique to playing the violin. The further you progress in a craft, vocation, sport, or instrument, the more gumption traps you'll discover—and the more interesting things will get. It's a sign of growing expertise.

If gumption is the psychic gasoline that fuels progress, then imagine a gauge that runs from full to empty. If you work hard, smart, and with integrity, you'll spend the majority of your time running near full. But every so often, the gauge falls and tensions rise. The above strategies add fuel to your tank and help you to continue on your path.

Chapter Summary

- Gumption is a forward inertia and enthusiasm that fuels progress.
- Adhere to the forty-eight-hour rule: Give yourself a set amount of time to revel in victory or wallow in defeat, but then get back to doing the work.
- See the ball go through the net to gain confidence and swagger.
- Step away from the work to refresh your enthusiasm when you are stuck.
- Explore outside of your domain for ideas and energy you can bring back to it.
- Surround yourself with people who will keep your ego in check.
- Stay flexible. What works, works great until it gets in the way.

Gumption is not about the absence of obstacles.

It's a feeling that no obstacle can stop you.

15

Curiosity

Prior to attempting a big lift in the gym, there's generally an element of fear. Will I make or miss the lift? What will a bar this heavy feel like? Am I willing, let alone able, to tolerate the discomfort? I often experience similar thoughts and feelings before speaking to a large crowd or shortly after committing to a challenging writing project. It's a mixture of doubt, fear, and *What am I getting myself into?*

Runners experience it on the start line of a race. Physicians feel it before performing a new procedure. Musicians feel it prior to a difficult solo. Discomfort is a natural by-product of pushing our limits—which is to say discomfort is a natural by-product of growth. When asked if he ever feels stage fright, Brian Wilson, the iconic leader of the Beach Boys, replied, "Oh, gosh yes. I have stage fright every concert."

Fear may be part of the process, but it doesn't mean we have to like it, and we certainly don't need to be paralyzed by it.

Whenever we'd approach the bar prior to one of those big lifts in the gym, my training partner began saying *brave new world*. It represents the fact that we don't know what is going to happen, but it sure will be interesting to see. I've come to embrace this mantra any time I'm exiting my comfort zone and advancing toward a new horizon. *Brave new world* shifts me from a tense, closed state of fear into a more playful and open state of curiosity. I become less *scared* about what could happen and more *interested* to find out. I feel and perform better as a consequence.

Herein lies the power of curiosity, a formidable antidote to fear that makes even the most challenging moments more satisfying. Curiosity is also integral to growth.

Think about how young children navigate their surroundings. They pick up objects and tinker, make all manner of sounds, try new movements, contort their bodies this way and that. When they stumble, fall down, or hit some other roadblock, they may experience brief frustration, but it's not long before they are exploring again. Nobody taught them to do this. They do it because learning is innately rewarding. Curiosity is in their nature—which means it is in our nature, too.

Our hardwiring for curiosity doesn't suddenly disappear when we reach a certain age. What happens is that our focus shifts to achieving specific outcomes, avoiding failure, and safeguarding our egos. Rather than perceiving challenges as opportunities to learn and grow, we view them as potential threats. Our fear of failure overrides the joy of exploration. We do everything possible to prioritize comfort and stability without realizing we're inhibiting our growth. We become nervous and scared instead of excited and curious.

Fortunately these behaviors can be unlearned. We can reconnect with our innate capacity for curiosity and harness it as we work toward mastery. A helpful starting point is understanding the tug-of-war between fear and exploration that happens inside our heads.

A Look Under the Hood

When confronted with significant uncertainty or challenge, a part of the brain called the amygdala livens with activity. The amygdala is an older structure, evolving early in the history of *Homo sapiens*. Its primary purpose is to trigger a stress response if we face grave danger, such as being attacked by a predator. It is a central part of what the late neuroscientist Jaak Panksepp coined the *PANIC* and *RAGE* pathways: the neural circuitry that is predictably activated when our sense of self comes under threat.

The PANIC and RAGE pathways evolved for good reason—if a tiger is chasing us, our survival depends upon our ability to feel fear and act on it. In the modern world, however, our threats rarely come from tigers. When pushing our limits and developing new skills, what we perceive as threats may not really be threats at all; and if they are, they are probably somewhat tolerable. Performing at our best is rarely a matter of life or death. Even when it is, as in the case of a trauma doctor, we improve our chances of getting into the zone and reaching peak states when we operate from a place of openness and curiosity rather than apprehension and fear. Thankfully we are equipped with neural pathways beyond PANIC and RAGE.

Another brain region, the basal ganglia, receives input from the amygdala via a group of neurons that make up a tiny structure called the striatum. You can think of the striatum as a bridge connecting the basil ganglia to the amygdala, as well as to other parts of the brain. The basal ganglia is not solely connected with PANIC or RAGE. It is implicated in other behaviors, too, including what Panksepp named the *SEEKING* and *PLAY* pathways.

The SEEKING pathway facilitates planning and problem-solving. It underlies our ability to exert agency and consciously move toward challenges instead of being overtaken by fear, descending into helplessness, or impulsively running away. The PLAY pathway is linked to exploration, discovery, and adventure. When we embrace a mindset of curiosity, we activate the SEEKING and PLAY pathways.

Panksepp's research shows that different pathways in the brain compete for resources, engaging in what amounts to a zero-sum game: If the SEEKING and PLAY pathways are activated, then the PANIC and RAGE pathways are deactivated.

Most of us can relate to this in our own lived experience. It is nearly impossible to be filled with curiosity and fear simultaneously. By engaging the features that make up the former, we prevent ourselves from spiraling into the latter. This explains why the mantra *brave new world* elicits such a stark shift in feelings.

What's more, the neural circuitry associated with curiosity is like a muscle: It gets stronger with use. That means if we are able to muster curiosity toward a challenge today, we'll be more likely to do so out of habit tomorrow. Curiosity is what neuroscientists call a *reward-based behavior*. It feels good, motivates us to keep going, and builds upon itself.

When we adopt curiosity as a default attitude toward challenges, we put ourselves in a better position to meet the moment and grow.

The Ultimate Reward Is Growth

Prior to his tragic death, the basketball legend Kobe Bryant was asked if he was a player who loves to win or hates to lose.

"I'm neither," he responded. "I play to figure things out. I play to learn something. If you play with a fear of failure or the will to win, I think it's a weakness either way. Because if you play with the fear of failing, you'll have the pressure on yourself to capitulate to that fear. If you play with the sense of I want to win, win, win, then you have the fear of what happens if you don't. But if you find common ground in the middle, in the center, then it doesn't matter. You aren't fazed by either. That enables you to really stay in the moment, to stay connected to it, and not feel anything other than what's in front of you."

Whether you play basketball or cello, repair cars or build tables, write books or coach young people—your craft can be a vessel for self-discovery. Think back to the early chapters of this book and the concept of *homeostatic upregulation.* It describes our biological imperative to flourish, evolve, and grow. There is no greater source of fulfillment and satisfaction. It's why walking the path of excellence is so rewarding. To create, grow, and express ourselves are all central to our nature.

"The real cycle you're working on is a cycle called yourself," writes Robert M. Pirsig of his experience repairing motorcycles. "The machine that appears to be 'out there' and the person that appears to be 'in here' are not two separate things. They grow toward Quality or fall away from Quality together."

Curiosity is intimately linked to growth.

There's a curiosity to figure out what you are capable of.

A curiosity to better know your craft.

A curiosity to better know yourself.

All are part and parcel of excellence.

Stuart McMillan, coach to over thirty-five Olympic medalists, has built a coaching philosophy around what he calls *safe-to-fail experiments*. "When I've got a problem I'm trying to solve with one of my athletes, when I'm trying to eke out a few one-hundredths of a second on a sprint, it's not like there is a lot of low-hanging fruit left," he tells me. "At that level, performance is so complex, and it's hard to isolate what variables might make a difference. The best I can do is run safe-to-fail experiments. I come up with various ways to change the training, and so long as the downside of failure isn't a bad injury or something else catastrophic, I run the experiments and see what we get. It's what makes coaching forever interesting and so much fun."

For someone like McMillan, a safe-to-fail experiment might mean adding an additional workout to a bobsledder's weekly plan, programming slightly more strength training for a 200-meter runner, or ever-so-slightly shifting the shin angle in a 100-meter start position. But we can run safe-to-fail experiments in just about anything.

If we are stuck or stagnating, instead of doubling down on what we've been doing (or spiraling into rumination), we can get curious and identify our own variables to adjust. It could mean changing the cadence of our work, the tools we use, the tempo of the melody, or something else altogether. The deeper we get on our respective paths of excellence, the more important it becomes for us to stay curious and get comfortable running safe-to-fail experiments. Attaining the last few decimal points of a marginal progress is slow and unpredictable. Where some people become bored and disinterested, others are curious and fascinated.

When the concert pianist Igor Levit was asked how he could bear even just listening to Beethoven's *Moonlight* Sonata, given how many times he's played it, he responded, "Yes. I played the sonata just recently. The more often I play a sonata, the more I work with it—the less I apprehend it, the more it eludes me, the happier I am with it, the more frequently I want to play it. I never say *I've got this. Next, please.* The goal is, I always want to arrive at the beginning."

Another way to strengthen the habit of curiosity is by carrying an analog notebook with you when you engage in your craft, or keeping a diary or journal. When you come across something particularly interesting—perhaps an unforeseen challenge, an aha moment, or a new feeling or experience—jot it down. Even if you never revisit your notes, the act of writing down these observations helps to foster curiosity toward what you are encountering. Your pursuit of excellence becomes a field study, one in which you are both a participant and an ethnographer.

The participant-self is playing the finite game, which is time bound and may include success or failure, winning or losing, a great day or a terrible day. The ethnographer-self is grounded in the infinite game, which knows no end. The only goal is to continue playing, learning, and growing. Excellence is an infinite game. The goal is the path and the path is the goal. So much of success simply comes down to staying on it.

Anxiety or Excitement? Nervous or Ready?

From the outside, chess appears to be a game that demands calm above all else. But that's not always the case, even and perhaps especially at the highest level. “Nerves are an essential part of the game,” the chess superstar Maxime Vachier-Lagrave told me. “If I am feeling too relaxed before a tournament, I cannot rise to the level of competition,” he said. “My nerves help me to focus and to fight.”

The superstar violinist Hilary Hahn loves performing, and she’s quite good at it, as evidenced by her two Grammys for Best Instrumental Soloist with an Orchestra. When I asked her what she feels prior to stepping onstage with a full ensemble at her back and thousands of fans filling the concert hall before her, she didn’t hesitate: “I am filled with adrenaline, and I am sharpened by it.”

Vachier-Lagrave and Hahn are demonstrating yet another way to practice curiosity, one that becomes second nature for experts in most fields. They pay close attention to the sensations in their bodies, and before they default to labeling a feeling as fear, they explore it at a more granular level. Is it tightness? Tingling? Moisture? Pain? Butterflies? Do the sensations remain static, or do they change from moment to moment? Instead of trying to push their embodied experience away, they become interested in it and use it to their advantage.

Research shows the greats don’t avoid nerves. They work *with* them. Elite performers feel the same jittery sensations and anxiety as everyone else; they just learn to take them along for the ride.

In a study including more than two hundred elite and non-elite swimmers, researchers used a psychological survey to measure stress before a major race. They also asked each athlete if they viewed stress as beneficial or harmful. Prior to the race, both the elite and non-elite swimmers experienced the same intensity of psychological and physiological arousal. The difference was that the non-elites viewed stress as something to avoid, ignore, or try to dampen. They felt stress would hurt their performance. The elites, on the other hand, got curious about what they were feeling in their bodies and interpreted their sensations as neutral, if not aiding their performance. What the non-elites immediately perceived as fear, the elites saw as their bodies preparing to rise to the occasion. As a result, the sensations didn't bother the elites nearly as much. They channeled their heightened arousal into explosiveness in the pool.

Additional research published in the *Journal of Experimental Psychology* shows that instead of trying to calm ourselves before consequential events, such as public speaking, it's better to "reappraise pre-performance anxiety as excitement." When we attempt to suppress pre-event nerves, we are inherently telling ourselves that something is wrong. Not only does this make the situation worse, but it also requires energy to fight the feeling of anxiety, energy that could be better spent on the task at hand. Rather than trying to push away what we are feeling, we're better off remaining curious and labeling our nerves as *excitement.* Doing so shifts us from a threat mindset (stressed and apprehensive) to an opportunity mindset (interested and ready to go).

Instead of thinking *I feel off, something must be wrong,* try thinking, *This is my nervous system rising to the occasion;* or *I am feeling this way because I care about what I am doing;* or *Once I get started, my training will take over and I'll be fine.* The worst way to make anxiety go away is by trying really hard to make anxiety go away. The best path forward usually involves taking the anxiety along for the ride, stepping into the arena, and getting reps under your belt. The more you accept and face the nerves, the more they fade. It's simple, but simple does not mean easy. It requires guts.

Anxiety isn't always the enemy. Sometimes it means that what you are attempting is hard, and that you care. Nothing is wrong with that. Caring is fuel. Learning to ride the waves is one of the most powerful skills you can develop, and like any other skill, it gets better with practice.

It's the difference between an internal dialogue of *How am I going to do this?* versus *brave new world*—and it's important whether we are facing a single acute challenge or the journey of excellence as a whole. In both cases, we feel and perform our best when we assume an attitude of curiosity and growth.

Chapter Summary

- Curiosity is a formidable antidote to fear.
- A growth mindset places curiosity at the forefront of our awareness—it means viewing the pursuit of excellence as a means for continuous development and self-exploration.
- In addition to the finite game, in which the goal is to win, there is also an infinite game, in which the goal is to keep playing and learning along the way.
- Safe-to-fail experiments are crucial to progress, especially as we become more skilled at our crafts.
- How we label and relate to the sensations we experience has a significant effect on how we feel and perform—as such, we want to get curious instead of shutting down, doing what we can to reframe anxiety as excitement.
- When faced with a challenge, think *brave new world.*

Excellence is an infinite game: The goal is the path, and the path is the goal.

Whether you play basketball or cello, repair cars or build tables, write books or coach teams, your craft can be a vessel for self-discovery.

16

Failure

Failure sucks. It is also inevitable. Keep going.

Plenty has been written on *failing well* and *failing forward,* but I think the above three sentences do more than 99 percent of the work.

Sure, we can grow and learn from failure, but there's no secret recipe.

As we come back from failure, we integrate its lessons into our identities. However, if we are going to experience growth and meaning, these attributes must come on their own schedule. The bigger and more challenging the failure, the longer it takes for us to feel good again. What doesn't kill us usually makes us stronger, but only with time.

I can sit here and write that you ought to expect failure, that any long-term path of excellence includes it, and that nobody achieves greatness without failing along the way. All of that is true. But in the moment, failure hurts. There's not much more to it, and that's okay.

There are many theories on resilience, but it really comes down to a few core components: leaning into community; allowing yourself to feel sadness and loss while working to maintain hope at the same time; being kind to yourself, patient, and persistent; establishing small routines that support your mental health; and committing to showing up and returning to the work as you trudge through the mess.

Remember, the things you care about break your heart. But you've got to trust the love of the pursuit is big enough to hold the hurt of failure.

Failure sucks. It is also inevitable.

Keep going.

17

Community

By all accounts, Caitlin Clark is one of, if not the, most dominant college basketball players to ever set foot on the planet. Nobody has scored more points or won more single-season awards. But perhaps the most memorable moment of Clark's collegiate career came in her junior season, after not a victory but a gutting championship loss to Louisiana State University. Following the game, Clark and her Iowa Hawkeyes teammates were devastated. They "bawled," recalls Clark. After showering and changing, the team went to a bar a few blocks away from their Dallas hotel.

The journalist Wright Thompson was embedded with the team. "The name of the bar was Happiest Hour, and the staff didn't seem prepared for two dozen very tired, very nostalgic, very thirsty women," Thompson writes in his ESPN story "Caitlin Clark and Iowa Find Peace in the Process."

Thompson goes on to explain that an Iowa fan asked Clark if he could buy the team a drink. "Twenty-two shots," Clark proclaimed, referring to the number she wears on her jersey. Soon after, a tray showed up with twenty-two shots on it, and the ladies indulged.

"That night might end up being Caitlin's favorite memory from college," writes Thompson. "This group of women truly loved one another and for the rest of their lives when they looked at their Final Four rings, or came to some anniversary and saw the banner hanging in the rafters, it is that love they would remember. And evenings like the one in Dallas after they lost the biggest game of their lives but still had one another."

Throughout this book, I've used the metaphor that striving for excellence is akin to climbing mountains. There are invigorating peaks to shoot for, but you spend nearly all of your time on the sides—waiting out bad weather, trudging forward with small and consistent steps, recovering from hard days, and celebrating good ones. Succeed or fail, the journey is almost always better when you have the right climbing partners. If you travel alone, you risk becoming isolated, resentful, and burning out. If you travel with others, you're likely to find joy, meaning, and purpose.

A paradox of human nature is wanting to be unique and independent on the one hand, while wanting to belong and be enmeshed in something larger on the other. One way to resolve this paradox is by striving for our own version of greatness in a community of others. The word *compete* comes from the Latin roots *com*, which means "together," and *petere*, which means to "seek," "rise up," or "strive." In its most genuine form, even competition is about rising together.

Social scientists have spent decades studying world-class organizations in diverse fields—places that produce numerous Olympians or groundbreaking scientific discoveries. Time and time again, they've found what sets the best apart from the rest isn't cutting-edge technology, ritzy facilities, or even standout individual performers. The real competitive advantage is a supportive community where everyone is dedicated to growth and lifting each other up. This sort of culture makes doing the hard thing a little easier, whether the hard thing is a specific task, maintaining an optimistic outlook during a string of setbacks, or grinding through a tedious stretch of work.

In researching for his book *Tribe,* the investigative reporter Sebastian Junger learned that many soldiers are more satisfied at war than at home. This finding doesn't make sense on its face, but Junger discovered that soldiers miss the strong sense of belonging and meaning they had at war. "Humans don't mind hardship, in fact, they thrive on it," he writes. "What they mind is not feeling necessary. Modern society has perfected the art of making people not feel necessary."

Mastery and mattering are core psychological needs.

Our ability to connect, communicate, and cooperate has long been our species' greatest asset, and it still is. As such, the pursuit of excellence is best when it includes other people—especially if they are the right ones.

Surround Yourself Wisely

For a landmark study involving the United States Air Force, psychologists tracked a cohort of cadets over four years. The researchers found that while there was significant variability in fitness gains and losses across all the cadets, there wasn't much variability within squadrons. Squadrons are groups of about thirty cadets to which an individual is randomly assigned prior to their first year.

Cadets spend the vast majority of their time interacting with peers in their squadron. In a sense, the squadron becomes a second family: Cadets in the same squadron eat, sleep, study, and workout together. Even though all the squadrons trained and recovered in the same manner, some squadrons showed vast increases in fitness over four years, whereas others did not.

The determining factor as to whether the thirty cadets within a squadron improved was the incoming fitness level (from high school) of the least fit person in the group. The effect of one's peers' physical fitness on their improvement was 70 percent as strong as the effect of their own fitness. In other words, if someone entering the Air Force Academy is highly fit and motivated to improve but they are assigned to a squadron with someone who is not, it nearly cancels out their own ability to make progress. "Even more strikingly," write the studies' authors, "the effect of the least fit peers on the probability of failing the [Air Force] fitness exam is larger than the effect of one's own incoming high school fitness." Just like diseases easily spread through tight-knit groups, so does performance—and it is highly contagious.

In another study, researchers at the University of Rochester had subjects watch a video of people describing themselves as being either intrinsically motivated or extrinsically motivated prior to playing a game. Those assigned to watch the video of people describing themselves as being intrinsically motivated reported feeling more intrinsically motivated themselves. Furthermore, when researchers left the subjects alone, those who had watched the intrinsically motivated video started playing the same game shown in the video of their own volition, while those who watched the extrinsically motivated video did not. Perhaps most fascinating is that these effects were strong irrespective of whether someone identified as being intrinsically or extrinsically motivated prior to the experiment. It's not just our fitness that pales in comparison to the fitness of those around us but also our attitudes.

Other research shows that if you sit within twenty-five feet of a high performer in a knowledge-work setting, your performance improves by 15 percent. Sit within twenty-five feet of a low performer, however, and it declines by 30 percent.

Emotions are contagious, too. Social scientists from Yale University monitored five thousand people living in the town of Framingham, Massachusetts, for more than three decades. They found that when someone became happy or sad, those feelings rippled throughout the entire community. A study aptly titled "I'm Sad, You're Sad" found that if you are in a negative mood when you text message your partner, they are likely to pick up on it and experience a lower mood state themselves. The same is true within online communities: Studies show that emotions spread like wildfire.

The point is not to ignore those experiencing negative emotions or who are down on their luck. Rather, we should muster all the compassion we can to support them. But there is a difference between a friend or teammate going through a hard time and someone who is perpetually negative, who lives to stoke anger and bring others down. Avoiding the latter is every bit as critical as supporting the former.

On the flip side, there is immense power in surrounding yourself with people who are committed to excellence. Community is advantageous if you are trying to improve at running, painting, writing, making music, teaching, or coaching. It is advantageous if you are a beginner, and it is advantageous if you are on the verge of becoming world-class.

When we think about progress and performance, we tend to emphasize the individual. But that's only half the story. Working to build a better self almost always means working to build a better community around it.

Communities of Practice

An uncomfortable truth is that the deeper your dedication to a specific pursuit, the more likely it is that you won't have much time or energy for people outside of it, at least not beyond your immediate family and closest friends. We simply cannot do it all. The cost of excellence includes trade-offs and sacrifices. While it's natural to drift from people when your interests and passions diverge, it cannot become an excuse to isolate and work yourself into the ground. It's why building community within your primary pursuit is so important.

Research shows the best way to forge meaningful relationships is through collectives. We tend to like people with whom we share interests and values, and there is less pressure to immediately hit it off when we become involved in a group versus trying to meet people one-on-one. Think of it as joining or developing *communities of practice*: groups of like-minded others who will support you on your journey and vice versa.

You could find a local gym to train at. Start a journal or book club. Sign up for a mastermind group or speakers series. Join a writers' workshop or a makers' space. You could go to the exhibitions of other local artists and invite them to yours. The options are endless. What they all have in common is a requirement of putting yourself out there and devoting time and energy to something beyond yourself. What you lose in individual freedom you gain manifold in belonging and mattering.

For my 2021 book, *The Practice of Groundedness*, I interviewed hundreds of people from different walks of life about what they value, how they spend their time, and their levels of fulfillment. I found that many view a tension between the effort required to build meaningful relationships and the type of optimization they believe is necessary for success. Time and time again, productivity gets prioritized over people. Even when we are exhausted and in need of renewal, the gold standard is *self-care,* in no small part because it is more efficient to do something alone than with others.

But the results of this mindset are anything but optimal. With this sort of heroic individualism, you can burn bright for a few days, weeks, or maybe even months. Yet over the long haul, what almost always happens is that you burn out.

Community is especially likely to get crowded out when we are on a roll. We become so engrossed in what we are doing that any diversion begins to feel like an inconvenience. This works fine—until we are confronted with our own vulnerability, be it an illness, injury, embarrassment, or failure. At that point, all of our optimizing leaves us alone and helpless.

If, however, we make it a priority to stay connected to a community of practice, what we'll get during setbacks is a pat on the back and some much-needed love. That's because everyone has undergone the same challenges. They know how hard it is. They *get it*. This shared understanding is priceless. It's why countless studies find that a central factor of resilience is social support.

We can rack up all the wins in the world, but if we don't have people to share them with, it's not long before the whole pursuit begins to feel empty.

Love

The Detroit Lions football team had just won their first playoff game in thirty-two years, breaking one of the worst slumps in the history of professional sports. Following the game, the scene in the locker room was one of jubilation. Massive men were singing, dancing, and crying tears of joy. During a short break from the raucousness, the head coach, general manager, and quarterback all gave brief speeches, which collectively lasted just over two minutes. Within those two minutes, the word *love* was repeated seven times. It ended with underdog quarterback Jared Goff struggling to hold it together, saying, "I love and appreciate you guys more than you know. I love you guys so much." The contrast between the scene in the Detroit Lions' locker room and the loneliness epidemic, which data show hits men particularly hard, could not be more striking. It's a prime example of why mastery and mattering are so vital, and why, regardless of gender, we must do everything we can to resist alienation and isolation by pursuing our own versions of excellence.

For the past seventy-five years, the Harvard Study of Adult Development has been tracking the physical and emotional well-being of over seven hundred men who grew up in Boston in the 1930s and 1940s. It later expanded to include their wives, sons, and daughters, too. It is one of the longest and most comprehensive longitudinal studies of its kind, closely following subjects from their late teens and early twenties all the way into their eighties and nineties. Many of the findings are what you'd expect: Don't drink too much, don't smoke, exercise often, eat a nutritious diet, maintain a healthy body weight, and stay intellectually engaged. Although according to George Vaillant, a psychiatrist and clinical therapist who directed the study for over three decades, the most important component of a long, happy, and healthy life is love.

"The seventy-five years and twenty million dollars spent on the Grant Study points to a straightforward five-word conclusion," he says. "Happiness is love. Full stop."

Many people think about love only in the romantic sense. But when the word originated in ancient Greece, it had multiple meanings. One of those was *philia,* which represented deep friendship, the brotherly and sisterly love that emerges out of shared struggles and triumphs. The Greeks valued this type of love as much as any. They understood how it enriches our lives with connection and meaning.

When we look back on our existence, we are less likely to focus on the gold medal, promotion to regional vice president, best-seller, or any other outward achievement than we are to cherish the bonds and relationships we forged along the way.

Unless you absolutely must, try not to go at it alone. The people with whom you surround yourself have an enormous impact on your pursuit of excellence and your life as a whole. In many ways, they shape it.

Chapter Summary

- Performance and motivation are contagious.
- Working to build a better self almost always means working to build a better community.

- Be wary of getting so caught up in self-driven optimization that you sacrifice belonging.
- Prioritizing community may feel inefficient in the short term, but in the long term, it is one of the most efficient things there is.
- Find or build *communities of practice*: groups of like-minded others who will support you on your journey and vice versa.
- Community supports you when you fall and provides gravity when you soar.
- "Happiness is love. Full stop."

We are all mirrors reflecting onto one another.

The people with whom we surround ourselves shape us.

18

Joy

In 2023, Courtney Dauwalter made history by becoming the first person to win ultra-running's triple crown in a single season: the Western States 100, Hardrock 100, and Ultra-Trail du Mont Blanc. That's three grueling mountain races, each a hundred miles or more, with a ridiculous amount of elevation, in a span of just ten weeks. It firmly planted her as one of, if not the, best ultra-runners in history. The running community believed such a feat was impossible until Dauwalter proved them wrong.

Yet Dauwalter wasn't out to *prove* anything to anyone. She wasn't even planning on running the third race. She essentially entered it on a whim, after surprising herself with how quickly she recovered following the second.

When my colleague Clay Skipper interviewed Dauwalter for our podcast, *excellence, actually*, he asked her what motivates her. She didn't hesitate: "Joy."

Dauwalter has always loved running, even before she became good enough to make a career out of it. "When I started running professionally, I wondered if that would take the joy out of it. But it hasn't. I love it even more probably. And I guess for me, I keep joy in the front seat of the car," she says. "Joy and intensity can coexist. But joy is driving the car. And if that's ever not true, then it will be a full reassessment of what's going on. Because I want it to stay fun, and I want it to be something I do my entire life. And if joy is driving, then the things I'm doing along the route are going to be because I love them, because they're fun, because they bring me that joy."

Does Dauwalter face pain and discomfort during races?

Absolutely.

There are times when she is cramping up alone on a ridgeline in the Rocky Mountains in the middle of the night at mile seventy-three of a race, knowing she has twenty-seven more to go. But it's because she experiences so much joy that she is able to endure and push through. It's her ultimate source of fuel. "No matter how bad I'm hurting or how much of a zombie I look like in the tough moments, I remind myself how lucky I am to be out there doing this thing I love," she says.

People love to romanticize the elite performer who has a chip on his shoulder, the person who is fueled by anger and who lives to prove naysayers wrong. Popular culture is full of individuals who exude an attitude of unwavering *hardness* for everything life throws their way. Though this approach plays well on the internet and can be genuinely effective in the short-term, it is not a viable strategy over the long haul.

If you are always angry, you won't have much fun. And if you don't have much fun, you probably won't last long in whatever it is you do.

There are a few glaring exceptions, the most famous being Michael Jordan. But it's worth pointing out that Jordan excelled most (and won all his championships) when being coached by Phil Jackson, the Zen devotee who was known for his calm and joyful approach to the game. Without Jackson to balance him out, it's easy to imagine Jordan would have self-destructed, like so many other athletes who fit his mold. And for all his ferocious intensity, even Jordan stuck his tongue out during dunks—a primal and perhaps involuntary expression of joy, freedom, and love for the game.

Relying predominantly on negative emotions for energy is akin to burning dirty fuel: It's destructive for the long-term health of the engine, and it's not sustainable. Joy, on the other hand, is a clean and renewable source of energy. It builds upon itself and doesn't leave toxic pollutants, such as resentment and isolation, in its wake.

Drive from Within

When Dauwalter left her job as a schoolteacher in 2017 to try running full time, she was an unknown entity. Just a few years later, she found herself on the cover of every major running magazine. Throughout it all, she's managed to keep joy in the driver's seat—not just during races but in everything else her meteoric rise in the sport has entailed.

It's easy to get caught up in hype, fame, and followers; in sponsorship deals, awards, and money. Chasing bright and shiny objects and external validation is seductive at first, but eventually it becomes exhausting. For one, this stuff is outside of our control. And no matter how much of it we get, it is never enough. It's like trying to quench our thirst with salt water—it only makes us thirstier. Intrinsic sources of drive, such as joy and mastery, are far more enduring.

"When you're younger and racing, for example, when you're running the mile in high school track, people ask you your time and your place immediately afterwards. And so you're taught and told that those are the things that you should value because those are the things that you're asked about right away," recalls Dauwalter. Yet, over time, she says that she's learned those aren't the real reasons for running. "It's about the shared moments and memories, not about all the other things and not about results and none of that," she explains.

Psychologists distinguish between two types of passion: harmonious and obsessive. With harmonious passion, you are absorbed in an activity because you love how the activity makes you feel. A harmoniously passionate artist creates because she finds joy in the craft. With obsessive passion, you get hooked on an activity because of external rewards and recognition. An obsessively passionate artist creates because she wants to boast about prestigious gallery shows and other accolades.

Decades of research show that people whose passion is predominantly obsessive lose touch with the joy intrinsic to their work. They become impatient, angry, and resentful, especially when things don't go well. Studies repeatedly find that obsessive passion is linked to burnout, anxiety, depression, and unethical conduct. Just think about any prominent story of fraud in business or doping in sport. Nearly all the main characters became addicted to the spotlight and driven by a combination of insecurity and anger. Harmonious passion, on the other hand, is undergirded by joy, curiosity, and a process mindset. It is associated with health, happiness, overall life satisfaction, and yes, performance, too.

Obsessive passion is a trap. Hardly anyone starts out with it, but as we develop an affinity for an activity and do it more often, we begin to experience positive results and receive praise, recognition, and rewards. Subtly, perhaps without even realizing it, we become more attached to the external validation we get from doing an activity than to the joy of doing the activity itself.

This trajectory is every bit as common as it is precarious. Even if we achieve legitimate success, if it is fed by a longing for external results, recognition, and rewards, then trouble lies ahead. That's because we'll always crave more. More money. More fame. More medals. More attention. More status. We get sucked into a vicious cycle of empty striving, and joy disappears.

To be sure, every athlete gets a jolt from winning, every craftsperson feels good when their work finds commercial success, and every salesperson loves closing a deal. It is also completely normal to feel angry when we are slighted or doubted. There's nothing wrong with experiencing these emotions. We just want to prevent them from becoming the *predominant* forces underlying our motivation. They can be in the backseat, or even in the passenger seat for short periods of time, but we must work to keep joy in the driver's seat.

Regularly reflecting on our overarching purpose can help. The more we remember why we got into our activity to begin with, the better. If that feels inaccessible, take a few days off and try again later. Something that may aid in the process is going on a hike, watching a sunset, or participating in another activity in nature.

The research of the University of California, Berkeley professor Dacher Keltner shows that experiencing natural awe helps us gain perspective and transition out of a narrow, frustrated, self-obsessed state and into a more expansive one.

Whenever I catch myself spending too much time harping on book sales, getting frustrated by the politics of publishing, or comparing myself to other authors, I've found disconnecting and spending time in nature, reading my favorite books, or listening to my favorite songwriters to be some of the best medicine. I almost always return to the work with a renewed sense of perspective and internal motivation. I am reminded that the foremost goal is to do good work for its own sake and to grow as a person.

This kind of perspective is enormously helpful.

(The result of which is that we produce better, more imaginative work.)

Intensity and drive are double-edged swords. They can be beautiful and enlarging when they are pointed toward growth and fueled by a love of craft. But they can become diminishing and destructive when they are pointed toward winning at all costs and fueled by anger or the need for approval.

After being pushed to expound upon her greatness, Dauwalter, who possesses the charismatic mixture of confidence and humility we discussed earlier, replied, "What I know is it's all really fun for me—like this pain cave exploration and running and trails and exploring with my feet. I'm always reminding myself how lucky I am to be doing this thing I love . . . When I decided to sign up for the third race of the triple crown, my attitude was *Let's have some fun and find out if it's possible to finish it.*"

She ended up winning by over forty minutes.

Savor the Journey

Imagine there are two climbers who equally desire to reach the peak of a mountain. The first climber is constantly thinking ahead to what it will feel like when he gets to the top. He's envisioning the pictures he'll take, the praise he'll receive from his peers, and how he'll finally attain self-worth. As such, he's in a rush to get where he's going. The second climber is more present for the journey. He's paying close attention to the experience along the way and remembering to stop and savor the views.

Which climber do you think is going to have a longer and more fulfilling career? Which climber do you think is more likely to feel restless and quit when he comes upon hard times?

Sustaining greatness requires the drive to keep pushing, but it also requires releasing from our tunnel vision on what lies ahead and learning to slow down and find joy where we are. It is not either-or. We need both of these qualities.

A large study found that athletes who focus on process rather than outcomes are more successful. It's why so many elite coaches, from Bill Walsh to Dawn Staley to Steve Kerr, embrace some variation of the same mentality: *Do the work, let the outcome take care of itself, and find joy in the process.* Research shows that an excessive focus on outcomes pushes us toward insecure striving (which makes people more susceptible to stress and anxiety), whereas those who find joy in the process perform from a place of security and freedom (and as a result, do better).

I told myself I'd celebrate once I wrote a bestseller. When I hit a major list, it wasn't nearly as fulfilling or joyful as I thought it would be. I thought I'd experience contentment when I made a five-hundred-pound deadlift. When I did, it took three minutes before I was thinking about what it would take to reach 550. In both instances, I fell for the arrival fallacy: I realized no outward achievement was going to fill whatever hole I had inside. The only thing that fills that hole is doing good work and loving good people—both of which are ongoing processes that must be reaffirmed again and again. So I was happy for a bit, and then I was quickly onto thinking about the next achievement.

"It's been presumed that when good things happen, people naturally feel joy for it," says Fred Bryant, a social psychologist whose work focuses on savoring. Yet his research suggests that we don't automatically respond to "good things" in ways that maximize their positive impact on our lives, something that is increasingly true in the rush and tumult of today's world. We quickly transition from one thing to the next, worried that if we don't move fast enough, we'll fall behind. But what ends up happening is that we miss out on so much joy along the way.

Bryant's work suggests that when something good happens, it is important to deliberately savor it. For example, instead of immediately proceeding to whatever is next, we could pause and take stock of the feelings we are experiencing in our body. Later on, we could share these experiences with our teammates, colleagues, and friends. We could also write about them in a diary or journal. When we relive joyful experiences, we further ingrain them in our memory, and thus have an easier time calling upon them for fortitude during trials and tribulations.

One of the most common regrets you hear from masters of craft toward the end of their careers: *I wish I would have stopped and enjoyed the good moments more.*

Keep this in mind on your own path.

When you're having a good day, give yourself permission to slow down, take a deep breath, and soak it up. Remember that the good days are the ultimate reward. In many ways, they are the point of the entire endeavor.

The best performers in the world are focused, determined, a little bit crazy, at times obsessive, and live mundane lifestyles that most people would find boring. That is all true. But the best performers in the world also experience deep joy in their crafts. What makes for greatness is being intense *and* joyful. It's the joy that makes the ferocious dedication, drive, and intensity sustainable.

If you find yourself losing joy, don't assume it has to be this way. Odds are, you're either moving too fast or getting too caught up in everything surrounding your craft at the expense of actually doing your craft itself. It's a common conundrum. Fortunately, it is also highly reversible.

You can begin rekindling joy right now.

Do everything you can to keep it in the driver's seat.

Chapter Summary

- Joy prevents burnout, promotes longevity, and offers a wellspring of resilience.

- Relying predominantly on negative emotions for energy is akin to burning dirty fuel. Joy, on the other hand, is a clean and renewable source of energy. It builds upon itself and doesn't leave toxic pollutants, such as resentment and isolation, in its wake.
- If you find yourself losing joy and getting caught up in obsession, come back to your overarching purpose for doing the work.
- We do not automatically reap the benefits of joy; rather, we must deliberately stop and savor the good moments.

Intensity and joy can coexist.

19

Completion

Excellence knows no end, but there are milestones along the way:

We finish the work. Compete at the event. Publish the book. Release the album. Each is an opportunity to learn and grow, to mark the passage of time and generate meaning. But it doesn't happen automatically. During my reporting for this book, of all the factors of excellence, when I asked people about moments of completion, the most common answer I received was some version of *You know, I probably need to be more thoughtful about that.*

In the mid 2020s, the philosopher Byung-Chul Han burst into the mainstream, in large part thanks to his observations on how we've lost touch with rituals and other ways to mark significant moments in our lives. As a result, we risk floating from one project or achievement to the next, all the while lacking concrete milestones that imbue our lives with meaning. The ultimate danger, Han argues, is that we float along without pause or reflection until suddenly we reach the end of our journey and have no idea how we got there. Our lives lose narrative value. "A life is no longer structured by sections, completions, thresholds, and transitions," writes Han. "Instead, there is a rush from one present to the next, an aging without growing old."

Throughout history, rites of passage and other meaning-making ceremonies were cornerstones of cultures that prided themselves on excellence. It's only more recently that we've left these rituals behind, allowing them to get crowded out by the tumult and busyness of everyday life. But there's no reason we can't reclaim them and make them our own.

Gregg Popovich of the NBA's San Antonio Spurs is one of the winningest, most revered coaches in the history of sports. Pop, as he's affectionally known, has a commanding presence. Though he comes off as intimidating, it's apparent to anyone who follows basketball how much he loves the game and how much he cares about his players. In addition to world-class strategizing, scouting, and developing schemes on the court, Pop's approach has at least one other hallmark: Following big wins, tough losses, and other significant milestones, he takes his team out for elaborate dinners. These dinners are big deals. The menus are prescreened, the seating assignments meticulously crafted. Restaurant staff are warned ahead of time to prepare for a lengthy feast.

ESPN has gone so far as to say Pop's dinners "have built the Spurs dynasty." As one former player explains, "To take the time to slow down and truly dine with someone in this day and age—I'm talking a two- or three-hour dinner—you naturally connect on a different level than just on the court or in the locker room."

The NBA season is long: eighty-two games, or more if you're

lucky enough to make the playoffs. The best players and coaches are always looking ahead. Whatever transpired in a specific game or practice quickly fades into the rearview so that preparation for the next can begin. Add in cross-country travel, media commitments, and the personal lives and obligations of players and coaches, and the entire season all too easily becomes a frantic whirlwind of activity.

Being in the NBA represents the pursuit of excellence at the highest level. But if you aren't careful, what ought to be an energizing and enlivening endeavor can gradually turn into an empty grind. The result for some players and coaches is a loss of meaning and love for the game. Pop's team dinners break up the season's perpetual forward motion, injecting much-needed gravity along the way.

"I haven't been a part of that anywhere else," the Spurs' Hall of Fame forward Pau Gasol explains. "And players know the importance of it as well."

ESPN writer Baxter Holmes has documented how the team dinners are archived in a scrapbook that resides in the San Antonio Spurs' complex. Pictures from the evening, a copy of the menu, and even labels from bottles of wine are included. Each entry marks the passage of time in a concrete and tangible way, and contributes to the mystique and meaning of a franchise known for excellence.

The Spurs team dinners take up a considerable amount of time and require extended planning. If they weren't so thoughtfully executed, they could easily be viewed as a distraction, as one more chore in an already long and jam-packed season. Instead, they've been crucial to the experience of the players and coaches, and to the organization's success.

There are numerous ways to define milestones. Sometimes they are obvious, such as races, recitals, and exhibitions. Other times we must create our own: What have we been working toward in a concerted manner? What feels like a natural point of completion? When does it make sense to turn the page on this phase of the work and begin the next?

It's wise to have at least one milestone per year. If you step back and evaluate your most important pursuits, you could probably pick it out right now. It's also wise to separate internal milestones from external ones, at least whenever possible. The day *you* complete a project is every bit as important as the day it is published, released, or judged by others.

The decisive achievement with excellence is that we got to where we are. But in order to reap the journey's full rewards, it's vital to pause and reflect along the way. What went well? What didn't go well? What am I proud of? What would I do differently next time? What relics or artifacts from this stage of the journey can I keep to remember all I've learned and become?

When we don't take time to pause and reflect, when we rush from one accomplishment to the next, we surrender gravity in both senses of the word: something that holds us to the path and something that enriches it with weight and significance.

Completion rituals come in many forms. It could be as simple as a night out at your favorite restaurant. Or writing three lessons you learned on a notecard, which you store in a special box. It could be as elaborate as gathering with colleagues or teammates for a retreat in the mountains. The details of the activity don't matter;

what does is stepping out of your day-to-day grind and marking important milestones and transitions.

Earlier, we discussed giving ourselves a set duration of time to revel in triumph following wins but then firmly nudging ourselves to return to the work. Whatever the timing is, make sure you truly stop to celebrate. Focus less on the successful outcome and more on the satisfaction you gained from giving something your all and having the chips fall your way. Bring to mind the challenges you've surmounted, the relationships you've forged, and the memories you've made.

When we commemorate specific moments on our path, we make our experiences concrete and give them opportunities to soak in. They become a part of who we are.

In the myth of Sisyphus, the gods condemn the protagonist to push a boulder up a hill, only to have it roll down each and every time, for all of eternity.

There are countless interpretations of the story. The one I've come to adopt is that we are all Sisyphus. We are all pushing a boulder up a hill, only to have it fall down—again and again and again.

Each of us plays our own eighty-two-game seasons. Excellence is about finding fulfillment in the struggle. It's about exerting ourselves with a smile on our face more often than a frown. It's about expressing our unique talents and gifts and creating joy, community, and meaning in the process.

Every time we reach the top of a mountain, it's important to pause and take stock of that particular ascent, knowing all the while that what comes next is yet another climb. By marking milestones and moments of completion, the broader journey, the one that never ends, becomes that much more satisfying.

Chapter Summary

- Completion rituals imbue our journeys with meaning and significance.
- If we don't have formal milestones, we can create our own, aiming for at least one every year.
- We must protect time and space to step outside the day-to-day grind for reflection.
- When we commemorate specific moments on our path, our experiences have opportunities to soak in—they become a part of who we are.

Experience plus reflection equals growth.

Conclusion: Reclaiming Excellence

In the opening of this book, I wrote that we are made to move toward excellence as a tree is made to move toward the sun. I hope I've helped you understand why. Nothing is more energizing and satisfying than giving our all to the activities we care about most. When we build a life around excellence, we reclaim our innate drive for self-expression, growth, and progress. We get the most out of ourselves and make a unique mark on the world.

Excellence is not perfection or winning at all costs. It is a process of becoming the best performer—and person—you can be. When you pursue goals that challenge you, put forth an honest effort, and endure highs, lows, and everything in between, you gain self-respect that nobody can take away.

I've done my best to define excellence, make the case for why it is central to a good life, and detail the core factors that give rise to it. I hope you've enjoyed reading and that you've found the text resonant. The best way to continue the conversation is by taking these ideas off the page and applying them in your life.

What does it look like for you to care deeply, to live with inner and outer quality? To set the right goals and keep the main things the main things? To cultivate focus, discipline, and gumption? To own your seat? To stay consistent, patient, and build community? To pour yourself into the big projects of your life while prioritizing rest and renewal, so that your efforts aren't mere flickers in time but sustainable and long-lasting? How can you push forward with intensity while holding on to joy and fulfillment?

These questions are for all of us. Answering them is the work of a lifetime.

Regardless of when you are reading this book, the future will be uncertain. The pace of change will accelerate. New technologies will emerge that have profound implications in every corner of our lives. But one thing will stay the same: the value of excellence.

Excellence is core to who we are.

When we reclaim excellence, we reclaim our humanity.

—Brad

Acknowledgments

Writing a book is, at root, a solitary endeavor. But throughout my career I've learned that the more I can make it a team effort, the better—for myself, and, more importantly, for the book.

Thanks are in order to the following people who have blessed me with their support, wisdom, and skill:

First and foremost, to my family: Caitlin, Theo, and Lila. You are the biggest room in my identity house, and I would not want it any other way. I love you all deeply.

To the Growth Equation: Steve Magness, Chris Douglas, Nate Mechler, and Clay Skipper. I am lucky to work with you all and am immensely proud of what we've built and the impact it is having on people's lives. We also have a lot of fun, and that's important, too. To my close friends and colleagues: Cal Newport and Zach Greenwald, who read and reread multiple sections of this book and then discussed with me how to make it better for hours and hours upon end; to Angie Treasure, for helping me figure out how to best share these ideas on the internet; to Mitch Black and Todd May, for reading early chapters; to James Meder, for spending hours with me working toward the perfect cover for the book; and to Anna Paustenbach, who helped me shape this book from the get-go.

To my literary agent, Laurie Abkemeier, for being the best coach, mentor, thought partner, and champion of my writing. You are excellent at what you do. I can't imagine my career without you.

To my wonderful editor, Nina Shield, for making this a better book, and for making me a better writer. To my publisher, Judith Curr, for continuing to believe in my work. To Emily Wichland, for turning over every word in her exacting copyediting. And to Daphney Guillaume, Aly Mostel, Ann Edwards, Louis Braverman, and the rest of the team at HarperOne, for doing everything possible to make this book excellent and get it into the hands of readers.

To all the extraordinary craftspeople, athletes, musicians, surgeons, creatives, coaches, and leaders who entrusted me with their stories.

To the late Robert M. Pirsig and all the other giants whose shoulders this book is built on. One of the greatest blessings and joys and sources of satisfaction in my life is to be in conversation with your work. I can only hope that my writing will impact others as your writing has impacted me.

To all my readers, without whom none of this would be possible, and who lift the ideas and language off the page and do their best to live it day in and day out.

Appendix 1: Guide to Application

As an ongoing reference, here are the key ideas and practices that fall under each factor of excellence. Think of it as a comprehensive guide, a menu of options to inform a practice that, over time, you'll make your own.

Excellence means *involved engagement* in *worthwhile pursuits* that support your *values and goals*. It combines *mastery* and *mattering*.

Care

- Passion doesn't strike like lightning; it develops over time.
- First comes fit, then comes grit.
- When you care about something deeply, you make yourself vulnerable; it is the cost of stepping into the arena, of having skin in the game.
- If you try to protect yourself by coasting or phoning it in, you miss out on the fullness and texture of life.
- Work to build an identity house that has more than one room.
- It takes courage to care.

Goals

- There is no greater illusion than thinking the accomplishment of some goal will change your life. What will change your life is who you become in the process of going for it.

- The best goals align with your values and push you ever so slightly outside of your comfort zone.
- SMART goals (specific, measurable, achievable, relevant, time-bound) work great for beginners, but as we become more advanced, we may benefit from goals that are broader and open-ended.
- Adopt a process-over-outcomes mindset: Set a goal, break it down into its incremental steps, and then focus on nailing those steps.
- Be the best at getting better—and remember that better isn't just about objective performance but also about becoming stronger, kinder, and wiser.

Trade-Offs

- Drop the weight of balance; it's an illusion.
- You can't do it all; figure out what matters and prioritize as best you can.
- Think about having different seasons of life for different pursuits.
- Too often we think a great feat was accomplished despite the approach being basic and uncomplicated, when in fact it was accomplished *because* the approach was basic and uncomplicated.
- Whenever possible, strive for simplicity: focus on the 99.9 percent and not the 0.1.
- Keep the main things the main things—in craft and life.

Focus

- We are at our best when devoting full attention to one task at a time, creating a sense of intimacy with whatever it is we are doing.
- Use *deep-focus blocks* for important activities.

- Digital devices fragment our attention even when they are turned off, face down, and not our own. As such, it is wise to remove them altogether. Out of sight is out of mind.
- The ideal window for attentive engagement ranges from about fifty minutes to two hours, followed by five-to-thirty-minute breaks.
- Focus is like a muscle: We need to train it.

Consistency

- Excellence does not come from intensity; it comes from consistency.
- Raise the floor: What you do on your bad days is every bit as important as what you do on your good days—don't freak out. Adjust your plan, give what you've got to give, and then move on. Don't let a bad day spiral into a bad week.
- When your North Star is consistency, you adopt a short memory, which ironically is key to playing the long game. You forget about what happened on a great day and get back to work. You forget about what happened on a bad day and get back to work.
- Do not become addicted to observable progress. Instead, frame the work as an ongoing practice; measure and judge your level of attention and effort; and let progress be a by-product of that.
- Become known for your consistency. When you win at consistency, you give yourself the best chance of winning at everything else. *Show up every day and give what you've got.*

Discipline

- A common misconception is that we need to feel good to get going, but often it's the opposite that is true: We need to get going to give ourselves a chance at feeling good.

- Discipline bridges the gap between motivation and action.
- Excellence requires setting boundaries and constraints, prioritizing positive freedom over negative freedom.
- Don't prejudge performance when you are feeling off. Just get started and give yourself a chance.
- Fierce self-discipline requires fierce self-kindness.

Renewal

- Rest and renewal are crucial to consistent progress: Stress plus rest equals growth.
- It takes discipline to rest, especially in a society that glorifies grinding, short-term gains, and pushing to extremes.
- Three defining features of restful activities:
  - You are not exerting self-control by *trying* to rest.
  - You are not engaging in activities that cause you stress or anxiety.
  - You are not turning rest into work by obsessively tracking your recovery.
- Plenty of activities meet the above criteria for rest, but a few in particular are universally beneficial: walking (or other light physical activity); hanging out with friends; spending time in nature; sleep; extended breaks, ranging from days off to vacations.
- Think about alternating between bursts of intense, deep-focus work and short breaks throughout the day. Sleep serves as a bridge between days. Extended breaks promote renewal and endurance over the long haul.
- Rest and renewal are not things we do at the expense of our primary pursuit; they are an integral part of our primary pursuit and our commitment to excellence.

Confidence

- Genuine confidence comes from evidence.
- *Self-efficacy* is the belief that you can show up, work through challenges, and excel when it matters most—you gain self-efficacy by putting in reps at whatever it is you do.
- Own your seat.
- Arrogance is loud and insecure. Confidence is quiet and humble.
- Trust your training.

Patience

- There is no such thing as an overnight breakthrough.
- To make a meaningful difference, the work you put in needs to persist long enough to break through inevitable barriers and plateaus.
- One small but powerful way to practice patience: Leave your phone or other digital device behind when you go about the usual activities of your day. It may seem trivial, but it helps to decondition our habituation to stimulation, novelty, and speed, and carries over into larger aspects of life.
- People overestimate what they can accomplish in a year but underestimate what they can accomplish in a decade.
- The accomplishment of micro-objectives primes us to persist.
- The longer we stay in the game, the more our surface area for luck increases.
- Progress over perfection: Good enough is often good enough, and good enough over and over again is how you end up with something great.

Routine

- Routines provide structure and predictability in an inherently complex and uncertain world, and prime us to perform at our best.
- There is no such thing as a magical routine: You've got to experiment and figure out what works for you.
- Do what you can to align the activities in your day with your chronotype.
- A simple yet powerful framework to lay the foundation for excellence is coming up with three daily, weekly, and monthly practices.
- Our adherence to routines ought to be rugged and flexible—we must be able to release from our routines when necessary, adjust if needed, and then come back to them as fast as we can.

Gumption

- Gumption is forward inertia, a sense of progress and possibility, and a strong yet measured enthusiasm that fuels the pursuit of progress.
- There are two categories of gumption traps: external (things outside of yourself) and internal (things within yourself).
- Adhere to your own version of the forty-eight-hour rule.
- See the ball go through the net.
- When you find yourself stuck, have the courage to step away from the work.
- Explore outside of your primary domain.
- Surround yourself with people who keep your ego in check.
- Stay flexible: What works, works, until it gets in your way.

Curiosity

- Curiosity is an antidote to fear.
- A growth mindset means viewing the pursuit of excellence as a means for continuous development and self-exploration.
- There is a finite game in which the goal is to win, and there is an infinite game in which the goal is to keep playing and learning along the way.
- How we label and relate to the sensations we experience has a significant impact on how we feel and perform (e.g., turning anxiety into excitement).
- When faced with a big challenge, we can use the mantra *brave new world* to shift us into a growth mindset and an attitude of curiosity.

Failure

- Failure sucks.
- Failure is inevitable.
- Keep going.

Community

- There is immense power in surrounding yourself with people walking the same or similar paths as you.
- Performance and motivation are contagious.
- Prioritizing community may feel inefficient in the short term, but if your goal is to sustain excellence over the course of a lifetime, then nothing is more important than nurturing your essential bonds.
- Do what you can to find or build *communities of practice*: groups of like-minded others who will support you on your journey and vice versa.

- When we prioritize community, what is lost in individual freedom is gained in belonging.
- "Happiness is love. Full stop."
- The people you surround yourself with have an enormous impact on your life. In many ways, they shape it.

Joy

- Joy prevents burnout, promotes longevity, and offers a wellspring of resilience.
- Keep joy in the driver's seat: "Joy and intensity can coexist. But joy is driving the car."
- Intrinsic motivation and harmonious passion are healthier and more enduring sources of energy than extrinsic motivation and obsessive passion.
- We do not automatically experience joy's benefits; rather, we must deliberately stop and savor the good moments.
- Joy is the ultimate source of endurance.

Completion

- Completion rituals imbue our journeys with meaning and significance, and ensure we don't float from one thing to the next without a sense of gravity.
- If we don't have formal milestones, we can create our own.
- When we commemorate specific moments on our path, our experiences have an opportunity to soak in—they become a part of who we are.

Appendix 2: Core Values

To help you determine your values, what follows is a list of seventy-five commonly held ones. It is not exhaustive but meant to prompt ideation. Begin by choosing somewhere between ten and fifteen values. Next, group like values together until you have three to five groups. For each group, come up with a single word that best captures the essence of what you are after. These are your values.

If you are stuck and don't know where to begin, think of people whom you respect and then ask yourself what it is about them that you admire. You could also imagine yourself ten, twenty, or maybe even thirty years down the road, looking back on current you. What qualities would older and wiser you be proud of? Both of these strategies are inroads to uncovering your values.

Recall the examples from chapter 5:

Health: Taking care of my body and mind so I can be around and functioning for the people and activities I care about.

Craft: Pursuing excellence in a select few endeavors that matter most to me.

Family: Being there for my kids and partner.

Community: Showing up for the important people and places in my life.

Truth: Living with integrity and telling it how it is.

Your values help you select goals and provide an anchor for all that you do.

- Achievement
- Adventure
- Appreciation
- Attentiveness
- Authenticity
- Authority
- Autonomy
- Balance
- Beauty
- Belonging
- Boldness
- Building
- Challenge
- Citizenship
- Community
- Compassion
- Competency
- Consistency
- Contribution
- Craft
- Creativity
- Curiosity
- Determination
- Diligence
- Discernment
- Discipline
- Drive
- Effectiveness
- Efficiency
- Empathy
- Excellence
- Fairness
- Friendship
- Fun
- Grit
- Growth
- Happiness
- Health
- Honesty
- Humility
- Humor
- Integrity
- Intellect
- Justice
- Kindness
- Knowledge
- Leadership
- Learning
- Love
- Loyalty
- Mastery
- Meaning
- Openness
- Optimism
- Patience
- Performance
- Persistence
- Poise
- Practice
- Quality
- Recognition
- Reputation
- Respect
- Responsibility
- Security
- Service

- Skillfulness
- Stability
- Success
- Sustainability
- Temperance
- Trust
- Truth
- Wealth
- Wisdom

Appendix 3: Sleep

In addition to what we covered in chapter 10 on renewal, here is a list of tips and practices to support sleep, consolidated from the world's leading sleep research:

- Ensure you expose yourself to natural (i.e., nonelectric) light throughout the day. This will help you maintain a healthy circadian rhythm.
- Exercise. Vigorous physical activity makes us tired, and when we are tired, we sleep. But don't exercise too close to bedtime, especially if you struggle with sleep.
- Limit caffeine intake, and phase it out completely five to six hours prior to your bedtime.
- Only use your bed for sleep and sex—not for eating, watching television, working on your laptop, or anything else. The one exception is reading a physical book prior to bed.
- Don't drink alcohol close to bedtime. Although alcohol can hasten the onset of sleep, it often disrupts the later and more important stages.
- Limit blue-light exposure in the evening, which tends to be most intense in our phones, computers, and tablet-like devices. If you must use these devices, turn them onto grayscale mode.

- Do not start working on hard, stressful activities—be they mental or physical—after dinner.
- If you struggle with a racing mind, try inserting a brief mindfulness meditation session prior to bed.
- When you feel yourself getting drowsy, don't fight it. Whatever you are doing can wait until the morning.
- Keep your room as dark as possible. If feasible, consider blackout blinds.
- Keep your room cool. Though there is individual variation, most people sleep best at a temperature that is a few degrees cooler than what you set your thermostat to during the day.
- Keep your smartphone out of the bedroom entirely. Not on silent. Out.
- The sleep supplement with the strongest evidence of efficacy is melatonin; it is most effective in situations that require you to shift your sleep schedule, such as when you are traveling across time zones or working nights. Talk with your physician to learn more.

Sources: National Sleep Foundation; University of Michigan Sleep Medicine

Notes

Introduction

8 *the costs are substantial:* Ben Wigert, "Employee Burnout Is the Biggest Myth," Gallup, March 13, 2020, https://www.gallup.com/workplace/288539/employee-burnout-biggest-myth.aspx.

8 *only 32 percent feel like they are fully engaged:* Jim Harter, "U.S. Employee Engagement Needs a Rebound in 2023," Gallup, January 25, 2023, https://www.gallup.com/workplace/468233/employee-engagement-needs-rebound-2023.aspx.

8 *Survey research from Statista shows:* Preeti Vankar, "Share of People Flourishing, Getting By, Languishing, and Struggling Worldwide," Statista, November 29, 2023, https://www.statista.com/statistics/1400661/share-of-people-flourishing-getting-by-languishing-struggling-worldwide.

Chapter 1: The Biology of Excellence

15 *"The people who weep before my pictures":* Aaron Rosen, "Leap of Faith—How Mark Rothko Reimagined Religious Art for the Modern Age," *Apollo*, August 29, 2020, https://www.apollo-magazine.com/rothko-chapel-houston-modernism-religion.

15 *"You feel it like hitting a baseball":* David Remnick, "The Scholar of Comedy," *New Yorker*, April 28, 2024, https://www.newyorker.com/culture/the-new-yorker-interview/the-scholar-of-comedy.

15 *Bacteria achieve their goal:* Hannah H. Tuson and Douglas B. Weibel, "Bacteria-Surface Interactions," *Soft Matter* 9, no. 17 (2013): 4368–80, https://doi.org/10.1039/C3SM27705D.

16 *"In their un-minded existence":* Antonio Damasio, *The Strange Order of Things: Life, Feeling, and the Making of Cultures* (New York: Pantheon Books, 2018), 20.

16 *From bacteria evolved:* Nicholas A. Lyons and Roberto Kolter, "On the Evolution of Bacterial Multicellularity," *Current Opinion in Microbiology* 24 (2015): 21–28, https://doi.org/10.1016/j.mib.2014.12.007.

17 *When this feeling occurred:* Benjamín Labatut, *The Maniac* (New York: Penguin Press, 2023).

20 *skill is "an adaptive, functional relationship":* Duarte Araújo and Keith Davids, "What Exactly Is Acquired During Skill Acquisition?" *Journal of Consciousness Studies* 18, nos. 3–4 (2011): 7–23, https://api.semanticscholar.org/CorpusID:142009512.

20 "Skill is not an act or action": Stu McMillan, interview by Brad Stulberg, September 26, 2024, Asheville, North Carolina.

21 *"when feel and real start intersecting":* "Feel vs Real by Tiger Woods," posted April 28, 2023, by Alexandre Laflamme GOLF COACH, YouTube, 58 secs., https://www.youtube.com/watch?v=9aLB9eDcB2g.

22 *"'As you wish'":* George Saunders, "A Book Is Made of Choices . . ." *Story Club with George Saunders*, Substack, May 12, 2024, https://georgesaunders.substack.com/p/a-book-is-made-of-choices?utm_campaign=email-post&r=737p&utm_source=substack&utm_medium=email.

22 *"When we focus on making a physical object":* Richard Sennett, "Crafting a New World," interview by Suzanne Ramljak, *Utne Reader*, December 14, 2009, https://www.utne.com/mind-and-body/crafting-a-new-world/.

23 *put forth the "somatic marker" theory:* Antonio R. Damasio, "The Somatic Marker Hypothesis and the Possible Functions of the Prefrontal Cortex," *Philosophical Transactions of the Royal Society B: Biological Sciences* 351, no. 1346 (1996): 1413–20, https://doi.org/10.1098/rstb.1996.0125.

24 *"touching the ecstatic and allowing it":* Rick Rubin, *The Creative Act: A Way of Being* (New York: Penguin Press, 2023), 230–33, 306.

Chapter 2: The Psychology of Excellence

29 *Thinking allows us to simulate:* Mark Solms, *The Hidden Spring: A Journey to the Source of Consciousness* (New York: W. W. Norton, 2022), 112.

30 *They called it* The Need for Chaos*:* Michael Bang Petersen, Mathias Osmundsen, and Kevin Arceneaux, "The 'Need for Chaos' and Motivations to Share Hostile Political Rumors," *PsyArXiv*, 2018, https://doi.org/10.31234/osf.io/6m4ts.

30 *A 2024 meta-analysis:* Kinga Bierwiaczonek, Sam Fluit, Tilmann von Soest, Matthew J. Hornsey, and Jonas R. Kunst, "Loneliness Trajectories Over Three Decades Are Associated with Conspiracist Worldviews in Midlife," *Nature Communications* 15 (April 29, 2024): 3629.

30 *Additional research shows that boredom:* Robert Brotherton and Silan Eser, "Bored to Fears: Boredom Proneness, Paranoia, and Conspiracy Theories," *Personality and Individual Differences* 80 (July 2015): 1–15.

32 belonging*, or a sense of connection:* Daniel T. Gilbert, "Social Psychological Aspects of the Self," in *Handbook of Theories of Social Psychology*, vol. 1, ed. Paul A. M. Van Lange, Arie W. Kruglanski, and E. Tory Higgins (London: SAGE Publications, 2011), 437, https://psikologi.unmuha.ac.id/wp-content/uploads/2020/02/SAGE-Social-Psychology-Program-Paul-A.-M.-Van

-Lange-Arie-W.-Kruglanski-E-Tory-Higgins-Handbook-of-Theories-of-Social-Psychology_-Volume-One-SAGE-Publications-Ltd-2011.pdf.

33 *"to cultivate these same qualities":* Peter Korn, *Why We Make Things and Why It Matters: The Education of a Craftsman* (Boston: David R. Godine, 2013), 46, 102.

34 *"I can't walk around": Court of Gold,* episode 2, directed by Jake Rogal (Netflix, 2025).

35 *"fully engaged in body and mind":* David Pizarro and Tamler Sommers, hosts, *Very Bad Wizards,* podcast, episode 224, "Hurts So Good (With Paul Bloom)," November 2, 2021, https://podcasts.apple.com/us/podcast/very-bad-wizards/id557975157?i=1000540517520.

35 *"It's a trick":* Pizarro and Sommers, *Very Bad Wizards.*

35 *engineering addictive flow experiences:* Natasha D. Schüll, "Review of *The Ethics of Invention* by Sheila Jasanoff," Natasha D. Schüll's Website, published June 2018.

35 *crucial difference between excellence and flow:* Mihaly Csikszentmihalyi, *Flow: The Psychology of Optimal Experience* (New York: Harper & Row, 1990), 49.

36 *"the highest good . . . the ultimate end":* Aristotle, *The Nicomachean Ethics,* trans. David Ross, ed. Lesley Brown (Oxford: Oxford Univ. Press, 2009).

36 *people report better in-the-moment feelings:* Veronika Huta and Richard M. Ryan, "Pursuing Pleasure or Virtue: The Differential and Overlapping Well-Being Benefits of Hedonic and Eudaimonic Motives," *Journal of Happiness Studies* 11, no. 6 (2010): 735–62, https://link.springer.com/article/10.1007/s10902-009-9171-4.

37 *"it had a peculiar synthetic, technological taste":* Robert M. Pirsig, *Lila: An Inquiry into Morals* (New York: Bantam Books, 1991), 283.

38 *"to learn to relax as quickly and completely as possible":* Hans Selye, *The Stress of Life* (New York: McGraw-Hill, 1956), 420.

39 *"Man must have some goal, some purpose":* Hans Selye, *Stress without Distress* (Philadelphia: Lippincott, 1974), 102.

39 *Studies show that our response to challenge:* Brad Stulberg, *Master of Change: How to Excel When Everything Is Changing—Including You* (New York: HarperOne, 2023).

39 *The neural activity associated with reward:* Mario Bogdanov, Héléna Renault, Sophia LoParco, Anna Weinberg, and A. Ross Otto, "Cognitive Effort Exertion Enhances Electrophysiological Responses to Rewarding Outcomes," *Cerebral Cortex* 32, no. 19 (2022): 4255–70, https://doi.org/10.1093/cercor/bhab480.

40 *The happiness and satisfaction he experienced as a by-product:* Hans Selye, *From Dream to Discovery: On Being a Scientist* (New York: Van Nostrand Reinhold, 1964), 9.

43 *invites a sitting response deep inside their brain:* Silvano Zipoli Caiani, "Extending the Notion of Affordance," *Phenomenology and the Cognitive Sciences* 13, no. 2 (2013): 275–93, https://link.springer.com/article/10.1007/s11097-013-9295-1.

43 *Mihaly Csikszentmihalyi, who coined the term* flow: Mihaly Csikszentmihalyi, *The Evolving Self: A Psychology for the Third Millennium* (New York: Harper Perennial, 1993), 139–41.

45 *"the defining element of feeling":* Antonio Damasio, *The Strange Order of Things: Life, Feeling, and the Making of Cultures* (New York: Pantheon Books, 2018), 102–6.

Chapter 3: The Philosophy of Excellence

49 *he continues to develop his theory of Quality:* Robert M. Pirsig, *On Quality: An Inquiry into Excellence* (New York: Bantam Books, 1991), 71.

49 *"Quality, selection, creates the world":* Pirsig, *On Quality*, 32, 72.

50 *de, which is translated as "virtue"*: Edward Slingerland, *Trying Not to Try: The Art and Science of Spontaneity* (New York: Crown, 2014), 7.

51 *When someone possesses de:* Slingerland, *Trying Not to Try*, 42.

51 *conatus symbolized an innate disposition:* Baruch Spinoza, *Ethics*, trans. by R. H. M. Elwes, 1677.

51 *"Goodness is really the fundamental":* Leo Tolstoy, *What Is Art?*, trans. Aylmer Maude (London: George Allen & Unwin, 1897), 77.

52 *"a particularly rewarding type of subjective experience":* Susan Wolf, *Meaning in Life and Why It Matters* (Princeton, NJ: Princeton Univ. Press, 2010), 27.

53 *associated with a range of negative:* Jalal Safipour, Donald Schopflocher, Gina Higginbottom, and Azita Emami, "The Mediating Role of Alienation in Self-Reported Health among Swedish Adolescents," *Vulnerable Groups & Inclusion* 2, no. 1 (2011): 5805, https://doi.org/10.3402/vgi.v2i0.5805.

54 *threw away his smartphone:* "Biography," John Moreland, https://johnmoreland.net/bio, accessed May 27, 2024.

57 *"skill is the greatest possession":* David Remnick, "The Scholar of Comedy," *New Yorker*, April 28, 2024, https://www.newyorker.com/culture/the-new-yorker-interview/the-scholar-of-comedy.

58 *"a lack of objective standards":* Matthew B. Crawford, *Shopclass as Soulcraft: An Inquiry into the Value of Work* (New York: Penguin Press, 2009), 15.

58 *"the car now runs, the lights are on":* Crawford, *Shopclass as Soulcraft*, 15.

59 *"place to improve the world":* Robert M. Pirsig, *Zen and the Art of Motorcycle Maintenance: An Inquiry into Values* (New York: William Morrow, 1974), 297.

Chapter 4: Care

68 *"bursts of high-impact work":* Lu Liu, Nima Dehmamy, Jillian Chown, C. Lee Giles, and Dashun Wang, "Understanding the Onset of Hot Streaks across Artistic, Cultural, and Scientific Careers," *Nature Communications* 12 (2021): 5392, https://www.nature.com/articles/s41467-021-25477-8.

69 *have greater longevity and achieve:* David Epstein, *Range: Why Generalists Triumph in a Specialized World* (New York: Riverhead Books, 2019).

71 *"the natural outcome of caring":* David Whyte, *Consolations: The Solace, Nourishment and Underlying Meaning of Everyday Words* (New York: Many Rivers Press, 2015), 101.

74 *we increase self-complexity:* Patricia W. Linville, "Self-Complexity and Affective Extremity: Don't Put All of Your Eggs in One Cognitive Basket," *Social Cognition* 3, no. 1 (1985): 94–120, https://doi.org/10.1521/soco.1985.3.1.94

Chapter 5: Goals

78 *"sides of the mountain which sustain life":* Robert M. Pirsig, *Zen and the Art of Motorcycle Maintenance: An Inquiry into Values* (New York: William Morrow, 1974), 204.

80 *91 percent of people fail:* "Why Most New Year's Resolutions Fail," *Lead Read Today* (blog), Fisher College of Business, Ohio State University, February 2, 2023, https://fisher.osu.edu/blogs/leadreadtoday/why-most-new-years-resolutions-fail.

82 *results in reduced motivation:* Maurice Schweitzer, Lisa Ordóñez, and Bambi Douma, "The Dark Side of Goal Setting: The Role of Goals in Motivating Unethical Behavior," *Academy of Management Proceedings* 2002, no. 1 (2002): B1–B6, https://doi.org/10.5465/APBPP.2002.7517522.

86 *"championship, or any accomplishment":* Ray Allen, *From the Outside: My Journey Through Life and the Game I Love* (New York: Houghton Mifflin Harcourt, 2018), 214.

86 the arrival fallacy: A.C. Shilton, "You Accomplished Something Great. So Now What?" *New York Times*, May 28, 2019, https://www.nytimes.com/2019/05/28/smarter-living/you-accomplished-something-great-so-now-what.html.

Chapter 6: Consistency

91 *sweet spot for accumulating workload:* Laura Bowen, Aleksander Stefan Gross, Mo Gimpel, and François-Xavier Li, "Accumulated Workloads and the Acute:Chronic Workload Ratio Relate to Injury Risk in Elite Youth Football Players," *British Journal of Sports Medicine* 51, no. 6 (2017): 452–59.

93 *a young Warren Buffett would tell:* Jason Zweig, "Warren Buffett and the

$300,000 Haircut," *Wall Street Journal*, July 5, 2019, https://www.wsj.com/articles/warren-buffett-and-the-300-000-haircut-11598626805.

97 *our bodies release the hormone dehydroepiandrosterone:* Andrew J. Elliot, ed., *Handbook of Approach and Avoidance Motivation* (New York: Psychology Press, 2008). Also see Brad Stulberg and Steve Magness, *Peak Performance: Elevate Your Game, Avoid Burnout, and Thrive with the New Science of Success* (New York: Rodale Books, 2017), 71.

101 *"When you have a long career like I have had"*: Mike Keegan, "Rory McIlroy insists he CAN end his 11-year major drought at the Masters—as he says he is ready to go through 'heartbreak' again at Augusta," *Daily Mail*, April 8, 2025, https://www.dailymail.co.uk/sport/golf/article-14584093/Rory-McIlroy-insists-end-11-year-major-drought-Masters-says-ready-heartbreak-Augusta.html.

Chapter 7: Trade-Offs

109 *strong internal self-awareness:* D. Scott Ridley, Paul A. Schutz, Robert S. Glanz, and Claire E. Weinstein, "Self-Regulated Learning: The Interactive Influence of Metacognitive Awareness and Goal-Setting," *Journal of Experimental Education* 60, no. 4 (1992): 293–306, http://www.jstor.org/stable/20152338.

109 *make better decisions:* Stephen L. Franzoi, Mark H. Davis, and Richard D. Young, "The Effects of Private Self-Consciousness and Perspective Taking on Satisfaction in Close Relationships," *Journal of Personality and Social Psychology* 48, no. 6 (1985): 1584–94, https://pubmed.ncbi.nlm.nih.gov/4020610/.

109 *have better personal relationships:* Paul J. Silvia and Maureen E. O'Brien, "Self-Awareness and Constructive Functioning: Revisiting 'the Human Dilemma,'" *Journal of Social and Clinical Psychology* 23 (2004): 475–89, https://doi.org/10.1521/jscp.23.4.475.40307.

109 *have more fulfilling careers:* Romila Singh and Jeffrey H. Greenhaus, "The Relation Between Career Decision-Making Strategies and Person–Job Fit: A Study of Job Changers," *Journal of Vocational Behavior* 64, no. 1 (2004): 198–221, https://doi.org/10.1016/S0001-8791(03)00034-4.

109 *improved mental health:* Robert C. Cloninger, "The Science of Well-Being: An Integrated Approach to Mental Health and Its Disorders," *World Psychiatry: Official Journal of the World Psychiatric Association* 5, no. 2 (2006): 71–76.

110 *results in decreased performance:* Robert J. Vallerand, "The Role of Passion in Sustainable Psychological Well-Being," *Psychological Well-Being* 2, no. 1 (2012): 1, https://doi.org/10.1186/2211–1522–2–1. Also see Brad Stulberg and Steve Magness, *The Passion Paradox: A Guide to Going All In, Finding Success, and Discovering the Benefits of an Unbalanced Life* (New York: Rodale Books, 2019).

Chapter 8: Focus

117 *As of 2023, economists estimated:* "New Economics for Sustainable Development: Attention Economy," United Nations Economist Network, February 2025, https://www.un.org/sites/un2.un.org/files/attention_economy_feb.pdf.

118 *our work suffers:* Sylvain Charron and Etienne Koechlin, "Divided Representation of Concurrent Goals in the Human Frontal Lobes," *Science* 328, no. 5976 (2010): 360–63, https://doi.org/10.1126/science.1183614.

118 *even seemingly innocuous multitasking:* Joshua S. Rubinstein, David E. Meyer, and Jeffrey E. Evans, "Executive Control of Cognitive Processes in Task Switching," *Journal of Experimental Psychology: Human Perception and Performance* 27, no. 4 (2001): 763–97, https://www.apa.org/pubs/journals/releases/xhp274763.pdf.

118 *A study conducted out of King's College:* "Multitasking Makes Us a Little Dumber," opinion, *Chicago Tribune*, August 10, 2010, https://www.chicagotribune.com/2010/08/10/multitasking-makes-us-a-little-dumber.

118 *"A wandering mind":* Steve Bradt, "Wandering Mind Not a Happy Mind," *Harvard Gazette*, November 11, 2010, https://news.harvard.edu/gazette/story/2010/11/wandering-mind-not-a-happy-mind.

118 *Multitasking detracts:* "Multitasking: Switching Costs," American Psychological Association, March 20, 2006, https://www.apa.org/topics/research/multitasking.

119 *average person checks their phone: Communications Market Report* (Ofcom, August 2, 2018), https://www.ofcom.org.uk/__data/assets/pdf_file/0022/117256/CMR-2018-narrative-report.pdf.

119 *Other research shows 71 percent:* "Americans Don't Want to Unplug from Phones While on Vacation, Despite Latest Digital Detox Trend," Asurion, December 5, 2018, https://www.asurion.com/press-releases/americans-dont-want-to-unplug-from-phones-while-on-vacation-despite-latest-digital-detox-trend.

119 *that number is forty-seven seconds:* Gloria Mark, "Should We Worry About Information Overload?" Substack, February 16, 2025, https://gloriamark.substack.com/p/should-we-worry-about-information.

121 *Still, the participants' performance declined:* Bill Thornton, Alyson Faires, Maija Robbins, and Eric Rollins, "The Mere Presence of a Cell Phone May Be Distracting: Implications for Attention and Task Performance," *Social Psychology* 45, no. 6 (2014): 479–88, https://doi.org/10.1027/1864-9335/a000216.

123 *ideal window for attentive engagement:* Brad Stulberg and Steve Magness, *Peak Performance: Elevate Your Game, Avoid Burnout, and Thrive with the New Science of Success* (New York: Rodale Books, 2017), 65.

125 *"great thoughts are conceived by walking":* Oliver Burkeman, "This Col-

umn Will Change Your Life: A Step in the Right Direction," *Guardian*, July 23, 2010, https://www.theguardian.com/lifeandstyle/2010/jul/24/change-your-life-walk-burkeman.

125 *walking break increased creative thinking:* Marily Oppezzo and Daniel L. Schwartz, "Give Your Ideas Some Legs: The Positive Effect of Walking on Creative Thinking," *Journal of Experimental Psychology: Learning, Memory, and Cognition* 40, no. 4 (2014): 1142–52, https://doi.org/10.1037/a0036577.

Chapter 9: Discipline

131 *Lewinsohn published his findings:* P. M. Lewinsohn, "A Behavioral Approach to Depression," in *The Psychology of Depression: Contemporary Theory and Research*, ed. R. J. Friedman and M. M. Katz (New York: John Wiley & Sons, 1974), 157–83.

131 *The more we try to suppress:* Daniel M. Wegner, David J. Schneider, Samuel R. Carter III, and Teri L. White, "Paradoxical Effects of Thought Suppression," *Journal of Personality and Social Psychology* 53, no. 1 (1987): 5–13, https://pubmed.ncbi.nlm.nih.gov/3612492/.

131 *When we try to force ourselves:* Jutta Joormann and Ian H. Gotlib, "Emotion Regulation in Depression: Relation to Cognitive Inhibition," *Cognition & Emotion* 24, no. 2 (2010): 281–98, https://doi.org/10.1080/02699930903407948.

132 *"Don't wait for the muse":* Stephen King, *On Writing: A Memoir of the Craft* (New York: Scribner, 2000).

133 *"the disciplined ones in life are free":* "Eliud Kipchoge & David Bedford, Full Address and Q&A," Oxford Union, posted January 5, 2018, by Oxford Union, YouTube, https://www.youtube.com/watch?v=Tc0OmDtzIJU.

133 *"discipline is freedom":* Claudiu Pop, "Venus Williams Reveals How She Fell in Love with Discipline in Mic Drop Speech," Tennis World, August 11, 2023, https://www.tennisworldusa.org/tennis/news/Tennis_Stories/136181/venus-williams-reveals-how-she-fell-in-love-with-discipline-in-mic-drop-speech.

134 *"If you're taking shortcuts":* Scott Cacciola, "Eliud Kipchoge Is the Greatest Marthoner Ever," *New York Times*, September 14, 2018, https://www.nytimes.com/2018/09/14/sports/eliud-kipchoge-marathon.html.

Chapter 10: Renewal

140 *"this discipline is acquired":* George Plimpton, "The Art of Fiction No. 21," interview with Ernest Hemingway, *Paris Review*, no. 18 (spring 1958), http://www.theparisreview.org/interviews/4825/the-art-of-fiction-no-21-ernest-hemingway.

141 *"A good idea doesn't come":* Jane E. Brody, "Hooked on Our Smartphones,"

New York Times, January 9, 2017, https://www.nytimes.com/2017/01/09/well/live/hooked-on-our-smartphones.html.

141 *"came to me on vacation":* Anna Almendrala, "Lin-Manuel Miranda Says It's No Accident 'Hamilton' Inspiration Struck on Vacation," *HuffPost,* June 23, 2016, https://www.huffpost.com/entry/lin-manuel-miranda-says-its-no-accident-hamilton-inspiration-struck-on-vacation_n_576c136ee4b0b489bb0ca7c2.

142 Stress plus rest equals growth: Brad Stulberg and Steve Magness, *Peak Performance: Elevate Your Game, Avoid Burnout, and Thrive with the New Science of Success* (New York: Rodale Books, 2017).

142 *Researchers frequently describe creativity:* Simone M. Ritter and Ap Dijksterhuis, "Creativity—the unconscious foundations of the incubation period," *Frontiers in Human Neuroscience* 8 (April 11, 2014): 215, doi:10.3389/fnhum.2014.00215.

143 *when athletes are on social media:* Petrus Gantois, Dalton de Lima-Junior, Leonardo de Sousa Fortes, Gilmario Ricarte Batista, Fabio Yuzo Nakamura, and Fabiano Fonseca, "Mental Fatigue From Smartphone Use Reduces Volume-Load in Resistance Training: A Randomized, Single-Blinded Cross-Over Study," *Perceptual and Motor Skills* 128, no. 4 (2021): 1640–59, doi:10.1177/00315125211016233.

144 *a reliance on sleep trackers:* Kelly Glazer Baron, Sabra Abbott, Nancy Jao, Natalie Manalo, and Rebecca Mullen, "Orthosomnia: Are Some Patients Taking the Quantified Self Too Far?" *Journal of Clinical Sleep Medicine* 13, no. 2 (2017): 351–54, doi:10.5664/jcsm.6472.

145 *"their sleep was suffering further":* Cody Gohl, "Study Finds Sleep Trackers May Make Insomnia Worse," *Sleepopolis,* June 2, 2021, https://sleepopolis.com/news/sleep-trackers-insomnia-worse-study.

145 *anchors of his daily routine:* Plimpton, "Art of Fiction No. 21."

145 *six minutes can produce benefits:* Marily Oppezzo and Daniel L. Schwartz, "Give Your Ideas Some Legs: The Positive Effect of Walking on Creative Thinking," *Journal of Experimental Psychology: Learning, Memory, and Cognition* 40, no. 4 (2014): 1142–52, https://aaalab.stanford.edu/assets/papers/2014/Give_your_ideas_some_legs.pdf.

146 *sport scientists call* active recovery*:* Olivier Dupuy, Wafa Douzi, Dimitri Theurot, Laurent Bosquet, and Benoit Dugué, "An Evidence-Based Approach for Choosing Post-Exercise Recovery Techniques to Reduce Markers of Muscle Damage, Soreness, Fatigue, and Inflammation: A Systematic Review with Meta-Analysis," *Frontiers in Physiology* 9 (April 2018): 403, doi:10.3389/fphys.2018.00403.

146 *social connection helps to shift:* Bethany E. Kok and Barbara L. Fredrickson, "Upward Spirals of the Heart: Autonomic Flexibility, as Indexed by Vagal Tone, Reciprocally and Prospectively Predicts Positive Emotions

and Social Connectedness," *Biological Psychology* 85, no. 3 (2010): 432–36, doi:10.1016/j.biopsycho.2010.09.005.

147 *who viewed pictures of nature:* Marc G. Berman, John Jonides, and Stephen Kaplan, "The Cognitive Benefits of Interacting with Nature," *Psychological Science* 19, no. 12 (2008): 1207–12, doi:10.1111/j.1467-9280.2008.02225.

148 *hardwired to feel at home:* Edward O. Wilson, *Biophilia* (Cambridge, MA: Harvard Univ. Press, 1984).

148 *lowers heart rate:* Yoshifumi Miyazaki, Juyoung Lee, Bum-Jin Park, Yuko Tsunetsugu, and Keiko Matsunaga, "[Preventive medical effects of nature therapy]," *Nihon eiseigaku zasshi* [Japanese journal of hygiene] 66, no. 4 (2011): 651–56, doi:10.1265/jjh.66.651.

149 *no such data exists:* "How Children Influence the Life Expectancy of Their Parents," Max Planck Institute for Demographic Research, October 23, 2019, https://www.mpg.de/14064449/children-influence-parents-life-expectancy.

150 *Ford introduced limits on work:* "Ford Factory Workers Get 40-Hour Week," *History*, last updated January 31, 2025, https://www.history.com/this-day-in-history/ford-factory-workers-get-40-hour-week.

150 *passed the Fair Labor Standards Act:* "Wages and the Fair Labor Standards Act," U.S. Department of Labor, https://www.dol.gov/agencies/whd/flsa, accessed July 24, 2024.

151 *reconnect with ourselves:* Carmen Binnewies, Sabine Sonnentag, and Eva J. Mojza, "Recovery During the Weekend and Fluctuations in Weekly Job Performance: A Week-Level Study Examining Intra-Individual Relationships," *Journal of Occupational and Organizational Psychology* 83, no. 2 (2010): 419–41, https://doi.org/10.1348/096317909X418049.

151 *Americans don't use all their paid time off:* Shradha Dinesh and Kim Parker, "More Than 4 in 10 U.S. Workers Don't Take All Their Paid Time Off," Pew Research Center, August 10, 2023, https://www.pewresearch.org/short-reads/2023/08/10/more-than-4-in-10-u-s-workers-dont-take-all-their-paid-time-off.

151 *"important to take a step back":* "009: Mental & Physical Skills of the World's Best (with Damian Warner)," *Farewell Podcast*, Apple Podcasts, February 1, 2024, https://podcasts.apple.com/us/podcast/009-mental-physical
-skills-of-the-worlds-best/id1505257676?i=1000643780403.

Chapter 11: Confidence

155 *"the confidence a farmer has":* Thich Nhat Hanh, *The Heart of the Buddha's Teaching: Transforming Suffering into Peace, Joy, and Liberation* (New York: Broadway Books, 1998), 185.

155 *"a matter of experimental faith":* Rick Rubin, *The Creative Act: A Way of Being* (New York: Penguin Press, 2023), 278–79.

157 *measures of self-efficacy:* Albert Bandura, "Self-Efficacy: Toward a Unifying Theory of Behavioral Change," *Psychological Review* 84, no. 2 (1977): 191–215, https://psycnet.apa.org/record/1977-25733-001.

157 *"And I am centered":* Maya Angelou, interview by Mavis Nicholson, *Mavis on 4,* February 9, 1987, posted February 10, 2017, by Thames TV, YouTube, https://www.youtube.com/watch?v=uZQJo0v_tlI.

158 *"mentors I could go to for advice":* Ryan Holiday, "Believing in Yourself Is Overrated. This Is Better," *RyanHoliday.net* (blog), November 22, 2016, https://ryanholiday.net/believing-in-yourself-is-overrated-this-is-better.

160 *"And that keeps you honest":* Angelou, interview by Mavis Nicholson, *Mavis on 4,* February 9, 1987.

161 *humility is associated with enhanced self-awareness:* Mark R. Leary, Kate J. Diebels, Erin K. Davisson, Katrina P. Jongman-Sereno, Jennifer Isherwood, Kaitlin T. Raimi, et al., "Cognitive and Interpersonal Features of Intellectual Humility," *Personality & Social Psychology Bulletin* 43, no. 6 (2017): 793–813, doi:10.1177/0146167217697695.

Chapter 12: Patience

163 *foundation of prior work:* Lu Liu, Yang Wang, Roberta Sinatra, C. Lee Giles, Chaoming Song, and Dashun Wang, "Hot Streaks in Artistic, Cultural, and Scientific Careers," *Nature* 559 (2018), 396–99, https://www.nature.com/articles/s41586-018-0315-8.

164 *"the love of science":* "Charles Darwin," NNDB: Tracking the Entire World, https://www.nndb.com/people/569/000024497, accessed June 10, 2020.

164 *online shoppers expected web pages to load:* "Akamai Reveals New Two-Second Web Page Response Threshold," Retail Technology, February 19, 2025, https://www.retailtechnology.co.uk/news/111/akamai-reveals-new-two-second-web-page-response-threshold.

164 *within a mere two-fifths of a second:* Steve Lohr, "For Impatient Web Users, an Eye Blink Is Just Too Long to Wait," *New York Times,* February 29, 2012, https://www.nytimes.com/2012/03/01/technology/impatient-web-users-flee-slow-loading-sites.html.

165 *"When moments without stimulation arise":* Teddy Wayne, "The End of Reflection," *New York Times,* June 11, 2016, https://www.nytimes.com/2016/06/12/fashion/internet-technology-phones-introspection.html?_r=0.

165 *"expectation for instant gratification":* Janna Anderson and Lee Rainie, "Millennials Will Benefit and Suffer Due to Their Hyperconnected Lives," Pew Research Center, February 29, 2012, https://www.pewresearch.org

/internet/2012/02/29/millennials-will-benefit-and-suffer-due-to-their-hyperconnected-lives/.

166 *"most of your time on a plateau":* George Leonard, *Mastery* (New York: Plume, 1992), 14–15.

167 *average age of a Nobel Prize–winning scientist:* R. Bjork, "The Age at Which Nobel Prize Research Is Conducted," *Scientometrics* 119, no. 2 (2019): 931–39, doi:10.1007/s11192-019-03065-4.

168 *peak until later in life:* "How Old Are Successful Tech Entrepreneurs?" *Kellogg Insight*, Kellogg School of Management, Northwestern University, May 15, 2018, https://insight.kellogg.northwestern.edu/article/younger-older-tech-entrepreneurs.

168 *"is in part a rejection of the role of experience":* Pierre Azoulay, Benjamin Jones, J. Daniel Kim, and Javier Miranda, "Age and High-Growth Entrepreneurship," Working Paper No. 24489 (National Bureau of Economic research, Cambridge, Massachusetts, April 2018), https://www.nber.org/system/files/working_papers/w24489/w24489.pdf.

170 *"accomplishes the great task":* Stephen Mitchell, *Tao Te Ching: A New English Version* (New York: HarperCollins, 1988), 63.

171 *process promotes progress:* Elliot T. Berkman, "The Neuroscience of Goals and Behavior Change," *Consulting Psychology Journal* 70, no. 1 (2018): 28–44, https://doi.org/10.1037/cpb0000094.

171 *"The science is good":* Gina Kolata, "Long Overlooked, Kati Karikó Helped Shield the World From the Coronavirus," *New York Times*, updated October 2, 2023, https://www.nytimes.com/2021/04/08/health/coronavirus-mrna-kariko.html.

173 *"Just a lucky day":* HT Sports Desk, "D Gukesh 'Shows True Meaning of Winning with Grace': '99 Out of 100 Times, I Would Lose to Magnus Carlsen. Lucky Day,'" *Hindustan Times*, last updated June 2, 2025, https://www.hindustantimes.com/sports/others/d-gukesh-shows-true-meaning-of-winning-with-grace-99-out-of-100-times-i-would-lose-to-magnus-carlsen-lucky-day-101748859465659.html.

173 *"Quality is a probabilistic function":* Dean Keith Simonton, "Talent and Its Development: An Emergenic and Epigenetic Model," *Psychological Review* 106, no. 3 (1999): 435–57, https://psycnet.apa.org/record/1999-03499-001.

173 *van Gogh created two of his most famous paintings:* Jessica Hallman, "Hot Streak: Finding Patterns in Creative Career Breakthroughs," PennState, September 6, 2018, https://news.psu.edu/story/535062/2018/09/06/research/hot-streak-finding-patterns-creative-career-breakthroughs.

174 *"when it's ready to love you back":* "Matt Campbell (Iowa State Head Football Coach) on Team Above Self & Falling in Love with the Process,"

speech, October 28, 2017, posted October 30, 2017, by Brandon Chambers, YouTube, https://www.youtube.com/watch?v=wuvnQ-yGjVk.

Chapter 13: Routine

177 *has nearly quadrupled in the last two decades:* "Google Books Ngram Viewer," Google Books, https://books.google.com/ngrams/graph?content=Morning+routine&year_start=1800&year_end=2019&corpus=en-2019&smoothing=3, accessed June 20, 2024.

178 *You settle into them:* George Leonard, *Mastery* (New York: Plume, 1992), 79.

179 *most people tend to perform at their best:* "Brain Function of Night Owls and Larks Differ, Study Suggests," *BBC News*, February 14, 2019, https://www.bbc.com/news/health-47238070.

180 *developed an evidence-based questionnaire:* J. A. Horne and O. Ostberg, "A Self-Assessment Questionnaire to Determine Morningness-Eveningness in Human Circadian Rhythms," *International Journal of Chronobiology* 4, no. 2 (1976): 97–110.

184 *"If you're lucky enough to be able to work from home":* "Astronaut Scott Kelly on Surviving Isolation: Dr. Sanjay Gupta's Coronavirus Podcast for April 3," CNN, April 3, 2020, https://www.cnn.com/2020/04/03/health/sanjay-gupta-podcast-transcript-april-3-wellness/index.html.

Chapter 14: Gumption

188 *it's a feeling that no obstacle can stop you:* Robert M. Pirsig, *Zen and the Art of Motorcycle Maintenance: An Inquiry into Values* (New York: William Morrow, 1974), 303–5.

189 *"can* keep *from getting fixed":* Pirsig, *Zen and the Art of Motorcycle Maintenance*, 303–5.

189 *"and leave you so discouraged you want to forget the whole business"*: Pirsig, *Zen and the Art of Motorcycle Maintenance*, 303–5.

192 *to run a mile in under four minutes:* Brad Stulberg and Steve Magness, *Peak Performance: Elevate Your Game, Avoid Burnout, and Thrive with the New Science of Success* (New York: Rodale, 2017), 91.

194 *"Some of the most creative breakthroughs occur":* Mihaly Csikszentmihalyi, *Creativity: Flow and the Psychology of Discovery and Invention* (New York: HarperCollins, 1996), 88–89.

Chapter 15: Curiosity

200 *When asked if he ever feels stage fright:* Brian Mackey, "Former Beach Boy Brian Wilson Reflects on Life and Music," *Norwich Bulletin*, October 14, 2009, https://www.norwichbulletin.com/story/news/2009/10/14/former-beach-boy-brian-wilson/46395561007.

203 *You can think of the striatum as a bridge:* José L. Lanciego, Natasha Luquin, and José A Obeso, "Functional Neuroanatomy of the Basal Ganglia," *Cold Spring Harbor Perspectives in Medicine* 2, no. 12 (December 1, 2012), https://doi.org/10.1101/cshperspect.a009621.

203 *It is implicated in other behaviors, too:* Joshua Kellman and Karam Radwan, "Towards an Expanded Neuroscientific Understanding of Social Play," *Neuroscience & Biobehavioral Reviews* 132 (2022): 884–91, https://doi.org/10.1016/j.neubiorev.2021.11.005.

203 *then the PANIC and RAGE pathways are deactivated:* Kenneth L. Davis and Christian Montag, "Selected Principles of Pankseppian Affective Neuroscience," *Frontiers in Neuroscience*, January 16, 2019, sec. *Neuropharmacology*, vol. 12 (2018), https://doi.org/10.3389/fnins.2018.01025.

203 *we'll be more likely to do so out of habit tomorrow:* Stephan J. Guyenet, *The Hungry Brain: Outsmarting the Instincts That Make Us Overeat* (New York: St. Martin's Press, 2017), 205.

203 *what neuroscientists call a* reward-based behavior*:* Andrew B. Barron, Eirik Søvik, and Jennifer L. Cornish, "The Roles of Dopamine and Related Compounds in Reward-Seeking Behavior Across Animal Phyla," *Frontiers in Behavioral Neuroscience*, October 11, 2010, sec. *Motivation and Reward*, vol. 4 (2010), https://doi.org/10.3389/fnbeh.2010.00163.

204 *"enables you to really stay in the moment"*: Jemele Hill, interview with Kobe Bryant, "Genius Talks: Kobe Unplugged," BET, July 3, 2015, https://www.youtube.com/watch?v=js8OfeEL4jI&t=70s.

204 *"They grow toward Quality or fall away from Quality together":* Robert M. Pirsig, *Zen and the Art of Motorcycle Maintenance: An Inquiry into Values* (New York: William Morrow, 1974), 325.

206 *When the concert pianist Igor Levit was asked:* Hartmut Rosa, *The Uncontrollability of the World* (New York: Polity, 2020), 45.

208 *the sensations didn't bother the elites:* Graham Jones, Sheldon Hanton, and Austin Swain, "Intensity and Interpretation of Anxiety Symptoms in Elite and Non-Elite Sports Performers," *Personality and Individual Differences* 17, no. 5 (1994): 657–63, https://www.sciencedirect.com/science/article/abs/pii/0191886994901384.

208 *Additional research published:* Alison Wood Brooks, "Get Excited: Reappraising Pre-Performance Anxiety as Excitement," *Journal of Experimental Psychology: General* 143, no. 3 (2014): 1144–58, https://doi.org/10.1037/a0035325.

Chapter 17: Community

215 *"This group of women truly loved one another":* Wright Thompson, "Caitlin Clark and Iowa find peace in the process," ESPN, March 20, 2024, https://www.espn.com/womens-college-basketball/story/_/id/39740282

/caitlin-clark-iowa-2024-ncaa-women-basketball-tournament-ready -march.

216 *The real competitive advantage is a supportive community:* Yafang Tsai, "Relationship between Organizational Culture, Leadership Behavior and Job Satisfaction," *BMC Health Services Research* 11, no. 98 (May 14, 2011), https://bmchealthservres.biomedcentral.com/articles/10.1186 /1472-6963-11-98.

216 *"the art of making people not feel necessary":* Sebastien Junger, *Tribe: On Homecoming and Belonging* (New York: Twelve, 2016), 17.

217 *"the effect of the least fit peers":* Scott E. Carrell, Mark Hoekstra, and James E. West, "Is Poor Fitness Contagious? Evidence from Randomly Assigned Friends," Working Paper No. 16518 (National Bureau of Economic Research, November 2010), https://doi.org/10.3386/w16518.

218 *those around us but also our attitudes:* Ron Friedman, Edward L. Deci, Andrew J. Elliot, Arien C. Moller, and Henk Aarts, "Motivational Synchronicity: Priming Motivational Orientations with Observations of Others' Behaviors," *Motivation and Emotion* 34, no. 1 (2010): https://psycnet.apa .org/record/2010-06261-004.

218 *your performance improves by 15 percent:* "Sitting Near a High-Performer Can Make You Better at Your Job," *Kellogg Insight*, Kellogg School of Management, Northwestern University, February 18, 2025, https://insight .kellogg.northwestern.edu/article/sitting-near-a-high-performer-can -make-you-better-at-your-job.

218 *those feelings rippled throughout the entire community:* James H. Fowler and Nicholas A. Christakis, "Dynamic Spread of Happiness in a Large Social Network: Longitudinal Analysis Over 20 Years in the Framingham Heart Study," *BMJ* 337 (2008): a2338, https://doi.org/10.1136/bmj.a2338.

218 *experience a lower mood state themselves:* Jeffrey T. Hancock, Kailyn Gee, Kevin Ciaccio, and Jennifer Mae-Hwah Lin, "I'm Sad You're Sad: Emotional Contagion in CMC," paper presented at proceedings of the annual ACM Conference on Computer Supported Cooperative Work, San Diego, CA, November 8–12, 2008, https://collablab.northwestern.edu/Collabolab Distro/nucmc/p295-hancock.pdf.

218 *emotions spread like wildfire:* Adam D. I. Kramer, Jamie E. Guillory, and Jeffrey T. Hancock, "Experimental Evidence of Massive-Scale Emotional Contagion Through Social Networks," *Proceedings of the National Academy of Sciences of the United States of America* 111, no. 24 (2014): 8788–90, https://doi.org/10.1073/pnas.1320040111.

221 *quarterback Jared Goff struggling to hold it together:* "Lions Win Wild Card Matchup vs. Rams: Locker Room Celebration | Extended Director's Cut," posted January 17, 2024, by Detroit Lions, YouTube, https://youtu.be /XTieOg0M17s.

222 *Many of the findings are what you'd expect: Adult Development Study,* "Publications," https://www.adultdevelopmentstudy.org/publications.

222 *"Happiness is love. Full stop":* George E. Vaillant, *Triumphs of Experience: The Men of the Harvard Grant Study* (Cambridge, MA: Belknap Press, 2015).

Chapter 18: Joy

229 *It is associated with health, happiness:* Robert J. Vallerand, "The Role of Passion in Sustainable Psychological Well-Being," *Psychology of Well-Being: Theory, Research and Practice* 2 (2012): 1, https://doi.org/10.1186/2211-1522-2-1.

230 *The research of the University of California, Berkely professor:* Dacher Keltner, *Awe: The New Science of Everyday Wonder and How It Can Transform Your Life* (New York: Penguin Press, 2023), 39.

231 *athletes who focus on process rather than outcomes:* Ollie Williamson, Christian Swann, Kyle Bennett, Matthew Bird, Scott Goddard, Matthew Schweickle, et al., "The Performance and Psychological Effects of Goal Setting in Sport: A Systematic Review and Meta-Analysis," *International Review of Sport and Exercise Psychology* 17, no. 2 (2022): 1050–78, https://doi.org/10.1080/1750984X.2022.2116723.

231 *an excessive focus on outcomes:* Vallerand, "The Role of Passion in Sustainable Psychological Well-Being," 1.

232 *"people naturally feel joy for it":* Stacey Kennelly, "10 Steps to Savoring the Good Things in Life," *Greater Good Magazine,* July 23, 2012, https://greatergood.berkeley.edu/article/item/10_steps_to_savoring_the_good_things_in_life.

Chapter 19: Completion

237 *"an aging without growing old":* Byung-Chul Han, *The Scent of Time,* trans. Daniel Steuer (Cambridge: Polity, 2017), 11.

237 *"you naturally connect on a different level":* Baxter Holmes, "Michelin Restaurants and Fabulous Wines: Inside the Secret Team Dinners That Have Built the Spurs' Dynasty," ESPN, July 25, 2020, https://www.espn.com/nba/story/_/id/26524600/secret-team-dinners-built-spurs-dynasty.